Ordinary Differential Equations

A Structured Approach

By Snehanshu Saha

University of Texas at El Paso

✿ cognella®
academic publishing

Bassim Hamadeh, CEO and Publisher
Michael Simpson, Vice President of Acquisitions
Jamie Giganti, Managing Editor
Jess Busch, Graphic Design Supervisor
Rose Tawy, Acquisitions Editor

First published in the United States of America in 2014 by Cognella, Inc.

Trademark Notice: Product or corporate names may be trademarks or registered trademarks, and are used only for identification and explanation without intent to infringe.

15 14 13 12 11   1 2 3 4 5

Printed in the United States of America

ISBN: 978-1-934269-704-8

cognella®
academic publishing

www.cognella.com   800-200-3908

To Pradosh Roy,
who inspired me to study Mathematics,
and to hundreds of students at The University of Texas at El Paso, who reaffirm my faith in the teaching profession every day.

Preface

There are plenty of books in Ordinary Differential Equations, many of which expertly written. The author had the privilege of studying some of those during his student days. Trying to teach out of one of those was a different experience. That's the time the thought of writing a book first popped up. Along came the offer from the publisher and here's the book: *Ordinary Differential Equations: A Structured Approach.*

Not everyone who chooses to learn differential equations or is required to take a class is a genius! A classic text may not help. There are plenty of well written books available that have loads of material clustered into one book. This is something the author wanted to circumvent. The book is written for an introductory, typical one semester course in Ordinary Differential Equations and doesn't tread a formal pedantic path.

The treatment is informal. It follows a simple language but covers the important techniques and concepts. In order to make the book self contained review chapters and a set of instructions in Mathematica are included in the book. The material in the book is just enough to be taught in one semester. No material is included for which the instructor doesn't have enough time to cover in class. That would defeat the purpose of the book.

Every effort is made so that someone could use it for self-teaching. This however should not deter any student from attending classes regularly.

A mere "thank you" would not suffice when it comes to Ruben Maguregui, undergraduate senior at The University of Texas at El Paso. He has been outstanding throughout the last year or so and immensely contributed toward the book.

I must be grateful to the publishers and sincerely thank Amy Whitbank, Kevin Fahey, and Rose Burt for being so kind and helpful. My colleagues at University of Texas at El Paso have been very encouraging. It would be an arduous task to single out anyone.

I thank my wife and family for the support and patience. I'm glad that my friends thought that I could pull off a mammoth task of writing a text.

Finally, a note of acknowledgement and appreciation to every single text book and its authors that I happened to read through is due.

Sincerely,

Snehanshu Saha,Ph.D.
The University of Texas at El Paso

Contents

Abbreviations Used in this Book

DE Differential Equation

ODE Ordinary Differential Equation

PDE Partial Differential Equation

IVP Initial Value Problem

BVP Boundary Value Problem

Chapter 0

Review of Calculus

Solving and understanding differential equations will rely on our familiarity with the ideas and techniques of differential and integral calculus. Though we will assume that the student has acquired such familiarity in previous mathematics courses, we will devote this chapter to a quick review of calculus, giving the student the opportunity to recall mathematical concepts and procedures that have to be mastered to more fluently understand the theory of differential equations.

0.1 Review of Differential Calculus

0.1.1 The concept of function

A function f from a set A to a set B can be understood as a rule that associates each element x in A to a unique element y in B. The set A is called the **domain of definition** of the function f, whereas the set B is called the **codomain** of f. The notation $f(x)$ is used to represent the element y in B associated to the element x in A. The element $f(x)$ is called the **value of f at the point x**, or the **image of x under the function f**. Recall that a function can be defined by means of an equation $y = f(x)$, in which the right-hand side contains an expression in terms of x only, which indicates how to obtain y for each of the values of x in the domain of the function. For example, the function f that associates each integer with its square can be defined by the equation

$$f(x) = x^2, \qquad x \in \{\ldots -3, -2, -1, 0, 1, 2, 3 \ldots\}.$$

In this case, the domain of f is the set of integers.

When the value of a variable y depends on the value of another variable x in such a way that the value of y corresponding to a value of x can be determined by a rule that is a function, we say that **y is a function of x**, and we call x and y the **independent variable** and the **dependent variable**, respectively. For example, the area of a circle, A, is a function of the radius r of the circle,

since

$$A(r) = \pi r^2, \qquad r \geq 0.$$

The restriction $r \geq 0$, which actually tells what the domain of this function is, comes from the fact that a circle cannot have a negative number as the length of its radius.

Let's see how we can define functions in *Mathematica*. We'll do this through an example:

Example 0.1

Define the function $f(x) = \cos x - \sin(2x)$ in *Mathematica* and then compute approximate values of $f(\pi)$, $f(-3)$ and $f(5.5)$. Use also the function **Plot** to produce the graph of this function in the interval $[0, 5\pi]$.

Solution: We will use the letter **f** to stand for the given function. The definition is then

In[1]:= **Clear[f]; f[x_] := Cos[x] - Sin[2 x]**

Out[1]:=

Now we compute $f(\pi)$, $f(-3)$ and $f(5.5)$, using the *Mathematica* function **N** to obtain approximations of each value.

In[2]:= **N[{f[π], f[-3], f[5.5]}]**

Out[2]:= {-1., -1.26941, 1.70866}

To plot a function y given by $y = f(x)$ in the interval $[a, b]$, use the function **Plot[f(x),{x,a,b}]**. Since we have already defined **f** to be **Cos[x] - Sin[2x]**, to get the graph of our function we simply write

In[3]:= **Plot[f[x], {x, 0, 5 π}]**

Out[3]:=

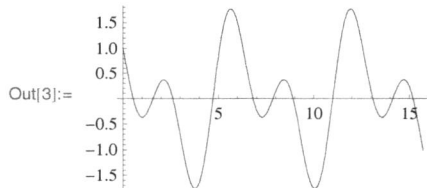

Sometimes y can be defined as a function of x in some interval (a, b) by means of an equation of the form $F(x, y) = 0$, in which the left-hand side contains an expression in terms of x and y. In this case we say that y is an **implicit function** of x. One way to test if an equation $F(x, y) = 0$ really defines y as a function of x is to solve the equation for the variable y, yielding an expression of the form $y = f(x)$, and we can easily check whether this new relation defines y as a function of x or not. However, solving for y in a given equation $F(x, y) = 0$ is in general a very difficult task, and determining analytically if the relation $F(x, y) = 0$ defines y as an implicit function of x requires knowledge of more advanced ideas from mathematical analysis. Nevertheless, we can often resort to the graph of $F(x, y) = 0$, from which we can obtain enough information to decide if this equation defines y as a function of x, and in which interval and under what restrictions this is so. The graph of $F(x, y) = 0$ can be obtained using one of several computer algebra systems, like *Mathematica*.

Example 0.2

Determine if the expression

$$x^2 + y^2 - 16 = 0$$

defines y as a function of x.

Solution: Solving the above equation for y is easy. Doing this yields the relation

$$y = \pm\sqrt{16 - x^2}.$$

Note that the square root in the right-hand side is not real unless $-4 \le x \le 4$, and hence y is defined only for values of x in the interval $[-4, 4]$. Moreover, the $\pm$ in front of the square root gives us two possible values of y for each value of x in $(-4, 4)$, one negative and one positive. Hence, the relation $x^2 + y^2 - 16 = 0$ *cannot*, by itself, define y as a function of x, for it does not associate a *unique* value y to each of the values of x. However, we can impose an additional restriction to this relation by choosing either the positive square root or the negative one. That is, we can choose one of the relations

$$y = \sqrt{16 - x^2} \qquad \text{or} \qquad y = -\sqrt{16 - x^2}$$

as the expression that defines y as a function of x, since with any of the two choices the original relation $x^2 + y^2 - 16 = 0$ is satisfied.

Example 0.3

Show that the relation

$$x^4 + x^2y^2 + y^4 - x(x^2 + y^2) = 0 \qquad\qquad (0.1)$$

defines a function y of x for all x in the interval $[0, 1]$.

Solution: In this case, solving for y is not easy, so we will inspect the graph of equation **(0.1)** and find some piece of it that corresponds to the graph of a function y of x. To plot the graph of **(0.1)**, we use the *Mathematica* function `ContourPlot`:

In[1]:= `ContourPlot[`$x^4 + x^2 y^2 + y^4 - x (x^2 + y^2)$ `== 0, {x, -0.1, 1.1},`
`{y, -1, 1}, PlotPoints → 100, Frame → None, Axes → True]`

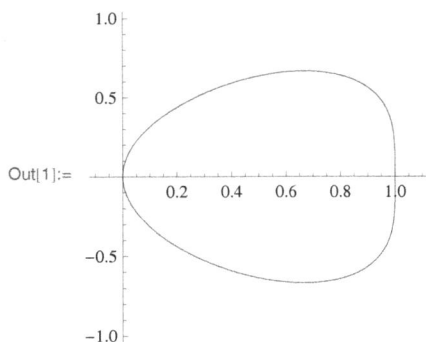

Out[1]:=

The plot above shows that we can select pieces of the graph of **(0.1)** that correspond to a function y of x. For example, we can select the section above the x-axis, that is, the section for which $y \geq 0$.

In[2]:= `ContourPlot[`$x^4 + x^2 y^2 + y^4 - x (x^2 + y^2)$ `== 0,`
`{x, -0.1, 1.1}, {y, -1, 1}, PlotPoints → 100, Frame → None,`
`Axes → True, RegionFunction → Function[{x, y}, y ≥ 0]]`

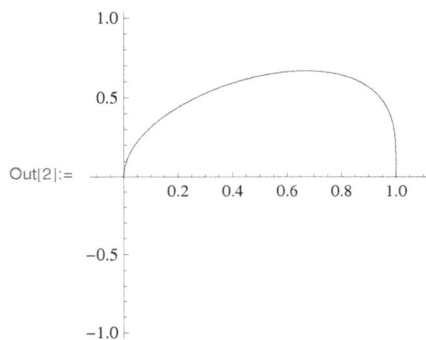

Out[2]:=

From this new plot, we see that the equation **(0.1)** does define y as a function of x for all $x \in [0, 1]$.

0.2   Continuity

Recall that a function $f(x)$ is **continuous** at a point a if and only if all of the following conditions are satisfied:

(1) $f(a)$ exists;

(2) $\lim_{x \to a}$ exists;

(3) $\lim_{x \to a} = f(a)$.

Example 0.4

Consider the three functions f, g and h given by

$$f(x) = \begin{cases} 2x & x \leq 1 \\ 1.5 & x > 1 \end{cases} \qquad g(x) = \frac{x^2 - 1}{x - 1} \qquad h(x) = \begin{cases} x + 2 & x \neq 1 \\ 1 & x = 1 \end{cases}$$

Show that none of these functions is continuous at 1.

Solution: The limit

$$\lim_{x \to 1} f(x)$$

does not exist, since

$$\lim_{x \to 1^+} f(x) \neq \lim_{x \to 1^-} f(x),$$

and therefore f cannot be continuous at 1. The function g is not continuous at 1 because $g(1)$ does not exist. For function h, both $h(1)$ and $\lim_{x \to 1} h(x)$ exist, but they are not equal:

$$h(1) = 1 \qquad \text{and} \qquad \lim_{x \to 1} = 2,$$

so h is not continuous at 1 either.

The example that follows illustrates the use of the option **Exclusions** for the **Plot** function in *Mathematica*, which is used to appropriately plot the graph of some discontinuous functions.

Example 0.5

Plot the graph of

$$f(x) = \frac{1}{x^2 - 1}$$

using the *Mathematica* function **Plot**.

Solution: If we use only the **Plot** function without defining some restrictions, we obtain

In[1]:= `Plot[`$\frac{1}{x^2-1}$`, {x, -2, 2}]`

Out[1]:=

Note the presence of the vertical lines passing through $x = -1, 1$, where we actually should not have any piece of graph at all. This can be corrected using the option **Exclusions** as follows.

In[2]:= **- GraphicsArray -**

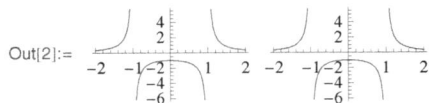

Out[2]:=

In the first plot of the above array, we have specified the values to be excluded explicitly, whereas in the second plot we asked *Mathematica* to figure out the points where the denominator is zero to exclude them.

The following functions are continuous everywhere in their domain.

(1) Polynomial functions: $p(x) = a_n x^n + a_{n-1} x^{n-1} + \cdots + a_1 x + a_0$, $n \in \mathbb{Z}$.

(2) Rational functions: $R(x) = \frac{p(x)}{q(x)}$, with p and q polynomial functions.

(3) Functions of the form $f(x) = \sqrt[n]{x}$ for n an odd integer. If n is an even integer, then $f(x) = \sqrt[n]{x}$ is continuous at each positive number.

(4) Logarithmic and exponential functions: $f(x) = \log_a x$ and $g(x) = a^x$, $a \in \mathbb{R}^+$.

(5) Trigonometric functions: $f(x) = \sin x$, $g(x) = \cos x$, etc.

We have the following fact.

Theorem 0.6
If $f(x)$ and $g(x)$ are continuous functions at a number a, then

(1) $f(x) + g(x)$ *is continuous at a;*

(2) $f(x) - g(x)$ *is continuous at a;*

(3) $f(x) \cdot g(x)$ *is continuous at a;*

(4) $f(x)/g(x)$ *is continuous at a provided $f(a) \neq 0$.*

0.2.1   Piecewise Functions

We will now introduce and briefly study some piecewise functions that are important in the study of differential equations. These functions are the **unit step function** $u(x)$, also called **Heaviside function**, and the **impulse function** δ_ϵ, given, respectively, by

$$H(x) = \begin{cases} 0 & t < 0 \\ 1 & t \geq 0 \end{cases} \quad \text{and} \quad \delta_\epsilon(x) = \begin{cases} \frac{1}{\epsilon} & 0 \leq x < \epsilon \\ 0 & x < 0 \text{ or } x \geq \epsilon. \end{cases}$$

Let's begin by studying the unit step function. First the graph of it:

In[3]:= `Plot[UnitStep[x], {x, -1, 1}, PlotStyle → Thick]`

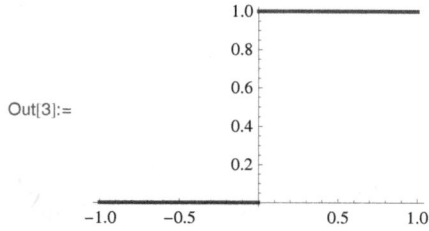

Out[3]:=

Note that `UnitStep` is the built-in *Mathematica* function for the unit step function u. To somewhat shorten our *Mathematica* code, let's make the following definition in this section.

In[4]:= `Clear[u]; u[x_] := UnitStep[x]`

Out[4]:=

Now we can play with some linear combinations of the step function, as for example

$$u(x), \qquad f(x) = \frac{1}{2}u(x-1) \qquad \text{and} \qquad g(x) = 2u\left(x+\frac{1}{2}\right) - 3u(x-1).$$

In[5]:= `f[x_] := `$\frac{u[x]}{2}$`; g[x_] := 2 u[x + `$\frac{1}{2}$`] + 3 u[x - 1]; (- GraphicsArray -)`

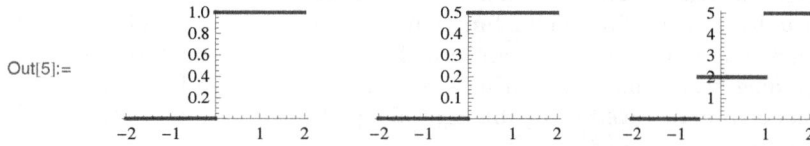

Out[5]:=

Note that any step function can be expressed as a linear combination of functions of the form $u(x-c)$. For example, the function f given by

$$f(x) = \begin{cases} 0 & x \leq -1 \\ 1 & -1 \leq x < 0 \\ -1 & 0 \leq x < 2 \\ 2 & x \geq 2 \end{cases}$$

can also be defined by $f(x) = u(x+1) - 2u(x) + 3u(x-2)$. In fact, a piecewise-defined function f defined by

$$f(x) = \begin{cases} f_1(x) & x < c \\ f_2(x) & x \geq c \end{cases}$$

is represented in terms of unit step functions as

$$f(x) = f_1(x) - u(x-c)f_1(x) + u(x-c)f_2(x).$$

For the impulse function δ_ϵ, we have

$$\delta_\epsilon(x) = \frac{1}{\epsilon}\left(u(x) - u(x - \epsilon)\right),$$

and its graph is shown below.

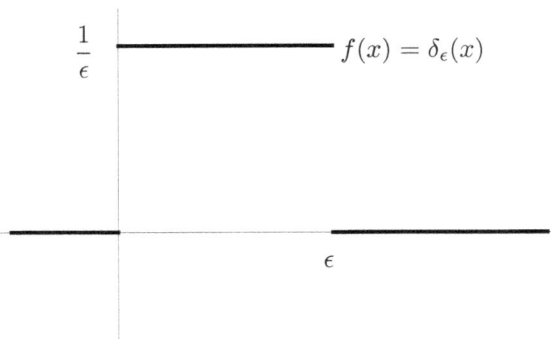

The **Dirac delta function** is defined as the limit

$$\delta(x) = \lim_{\epsilon \to 0^+} \delta_\epsilon(x) = \begin{cases} \infty & x = 0 \\ 0 & x \neq 0. \end{cases}$$

Strictly speaking, the Dirac delta function is not a function, but an object called a **distribution**. The Dirac delta function plays an important role in the study of phenomena in which there is an **impulse force**, which is a force of a relatively large magnitude that acts during a relatively short period of time. These phenomena can be modeled using differential equations that involve the Dirac delta function.

0.3 Differentiation

For reference, we will start with a list of differentiation rules. We expect the student to be familiar with these rules and their use.

(4) $\dfrac{d}{dx}(u+v) = \dfrac{d}{dx}u + \dfrac{d}{dx}v.$

(5) $\dfrac{d}{dx}(uv) = u\dfrac{d}{dx}v + v\dfrac{d}{dx}u.$

(6) $\dfrac{d}{dx}\left(\dfrac{u}{v}\right) = \dfrac{v\frac{d}{dx}u - u\frac{d}{dx}v}{v^2}.$

(7) $\dfrac{d}{dx}(u^r) = ru^{r-1}\dfrac{d}{dx}u.$

(8) $\dfrac{d}{dx}e^u = e^u\dfrac{d}{dx}u.$

(9) $\dfrac{d}{dx}a^u = a^u\ln a\dfrac{d}{dx}u.$

(10) $\dfrac{d}{dx}\ln u = \dfrac{1}{u}\dfrac{d}{dx}u.$

(11) $\dfrac{d}{dx}\sin u = \cos u\dfrac{d}{dx}u.$

(12) $\dfrac{d}{dx}\cos u = -\sin u\dfrac{d}{dx}u.$

(13) $\dfrac{d}{dx}\tan u = \sec^2 u\dfrac{d}{dx}u.$

(14) $\dfrac{d}{dx}\cot u = -\csc^2 u\dfrac{d}{dx}u.$

(15) $\dfrac{d}{dx}\sec u = \sec u\tan u\dfrac{d}{dx}u.$

(16) $\dfrac{d}{dx}\csc u = -\csc u\cot u\dfrac{d}{dx}u.$

(14) $\dfrac{d}{dx}\arcsin u = \dfrac{1}{\sqrt{1-u^2}}\dfrac{d}{dx}u.$

(15) $\dfrac{d}{dx}\arccos u = -\dfrac{1}{\sqrt{1-u^2}}\dfrac{d}{dx}u.$

(16) $\dfrac{d}{dx}\arctan u = \dfrac{1}{1+u^2}\dfrac{d}{dx}u.$

(17) $\dfrac{d}{dx}\operatorname{arccot} u = -\dfrac{1}{1+u^2}\dfrac{d}{dx}u.$

(18) $\dfrac{d}{dx}\operatorname{arcsec} u = \dfrac{1}{u\sqrt{u^2-1}}\dfrac{d}{dx}u.$

(19) $\dfrac{d}{dx}\operatorname{arccsc} u = -\dfrac{1}{u\sqrt{u^2-1}}\dfrac{d}{dx}u.$

(20) $\dfrac{d}{dx}\sinh u = \cosh u\dfrac{d}{dx}u.$

(21) $\dfrac{d}{dx}\cosh u = \sinh u\dfrac{d}{dx}u.$

(22) $\dfrac{d}{dx}\tanh u = \operatorname{sech}^2 u\dfrac{d}{dx}u.$

(23) $\dfrac{d}{dx}\coth u = -\operatorname{csch}^2 u\dfrac{d}{dx}u.$

(24) $\dfrac{d}{dx}\operatorname{sech} u = -\operatorname{sech} u\tanh u\dfrac{d}{dx}u.$

(25) $\dfrac{d}{dx}\operatorname{csch} u = -\operatorname{csch} u\coth u\dfrac{d}{dx}u.$

Another important differentiation rule is the **chain rule**: if f and g are two functions, with g differentiable at x and f differentiable at $g(x)$, then $f \circ g$ is differentiable at x and

$$\frac{d}{dx}(f \circ g) = \left(\frac{df}{dx} \circ g\right)\frac{dg}{dx}.$$

For this rule we will show an example:

Example 0.7

Let f and g be the functions given by

$$G(x) = (x^2 + 2)^5 \qquad \text{and} \qquad h(x) = \sin(\cos x),$$

compute $f'(x)$ and $g'(x)$.

Solution: Note that if we define $g(x) = x^2 + 2$ and $f(x) = x^5$, then $G(x) = (f \circ g)(x)$, and by the chain rule,

$$G'(x) = f'(g(x))g'(x) = 5(x^2 + 2)^4(2x) = 10x(x^2 + 2).$$

Differentiation is implemented in *Mathematica* by the function **D**. Here are some examples:

In[1]:= **Clear[f, g];** $\left\{\partial_t \, (\text{Cos}[t] \, \text{Sin}[t]), \, \partial_x \, (x+1)^5, \right.$
    $\left. \partial_x \left(ax^2 + bx + c\right), \, \partial_x \, f[g[x]], \, \partial_{g[x]} \, f[g[x]] \right\}$

Out[1]:= $\left\{\text{Cos}[t]^2 - \text{Sin}[t]^2, \, 5 \, (1+x)^4, \, 0, \, f'[g[x]] \, g'[x], \, f'[g[x]] \right\}$

Note from the last two differentiations that *Mathematica* is able to apply the chain rule when computing the derivative of a function.

0.4   Integral Calculus

In this section we will review common techniques to evaluate integrals. Let us begin with a typical list of rules to evaluate simple integrals.

(1) $\int a \, du = au + C$, with a any constant.

(2) $\int [f(u) + g(u)] \, du = \int f(u) \, du + \int g(u) \, du$

(3) $\int u^n \, du = \frac{u^{n+1}}{n+1} + \text{C}, \, n \neq -1$.

(4) $\int \frac{1}{u} \, du = \ln|u| + C$

(5) $\int a^u \, du = \frac{a^u}{\ln a} + C$, where $a > 0$ and $a \neq 1$.

(6) $\int e^u \, du = e^u + C$

(7) $\int \sin u \, du = -\cos u + C$

(8) $\int \cos u \, du = \sin u + C$

(9) $\int \sec^2 u \, du = \tan u + C$

(10) $\int \csc^2 u \, du = -\cot u + C$

(11) $\int \sec u \tan u \, du = \sec u + C$

(12) $\int \csc u \cot u \, du = -\csc u + C$

(13) $\int \tan u \, du = \ln|\sec u| + C$

(14) $\int \cot u \, du = \ln|\sin u| + C$

(15) $\int \sec u \, du = \ln|\sec u + \tan u| + C$

(16) $\int \csc u \, du = \ln|\csc u - \cot u| + C$

(17) $\int \frac{1}{\sqrt{a^2 - u^2}} \, du = \arcsin \frac{u}{a} + C$

(18) $\int \frac{1}{a^2+u^2}\,du = \frac{1}{a}\arctan\frac{u}{a} + C$

(19) $\int \frac{1}{u\sqrt{u^2-a^2}}\,du = \frac{1}{a}\operatorname{arcsec}\frac{u}{a} + C$

(20) $\int \sinh u\,du = \cosh u + C$

(21) $\int \cosh u\,du = \sinh u + C$

(22) $\int \operatorname{sech}^2 u\,du = \tanh u + C$

(23) $\int \operatorname{csch}^2 u\,du = -\coth u + C$

(24) $\int \operatorname{sech} u \tanh u\,du = -\operatorname{sech} u + C$

(25) $\int \operatorname{csch} u \coth u\,du = -\operatorname{csch} u + C$

A very common technique for evaluating integrals is that of **integration by parts**. If f and g are two differentiable functions, then

$$\int f(x)g'(x)\,dx = f(x)g(x) - \int g(x)f'(x)\,dx$$

If we define $u = f(x)$ and $v = g(x)$, then $du = f'(x)\,dx$ and $dv = g'(x)\,dv$, and the above formula can be written as

$$\int u\,dv = uv - \int v\,du. \tag{0.2}$$

Let us look at some examples.

Example 0.8

Evaluate

$$\int x \ln x\,dx.$$

Solution: To apply the above formula, we define $u = \ln x$ and $dv = x\,dx$, so $du = \frac{1}{x}\,dx$ and $v = \int x\,dx = \frac{1}{2}x^2$. Hence

$$\begin{aligned}
\int x \ln x\,dx &= \frac{1}{2}x^2 \ln x - \int \frac{1}{2}x^2\left(\frac{1}{x}\right)\,dx \\
&= \frac{1}{2}x^2 \ln x - \frac{1}{2}\int x\,dx \\
&= \frac{1}{2}x^2 \ln x - \frac{1}{4}x^2 + C.
\end{aligned}$$

Example 0.9

Use integration by parts to evaluate the integral

$$\int e^{at} \sin(bt) \, dt.$$

Solution: Define $u = \sin(bt) \, dt$ and $dv = e^{at}$. Therefore, $du = b\cos(bt)$ and $v = \frac{1}{a}e^{at}$, so

$$\int e^{at} \sin(bt) \, dt = \frac{1}{a}e^{at} \sin(bt) - \frac{b}{a} \int e^{at} \cos(bt).$$

Now we apply integration by parts to the integral in the RHS of the above expression:

$$\int e^{at} \cos(bt) = \frac{1}{a}e^{at} \cos(bt) + \frac{b}{a} \int e^{at} \sin(bt).$$

Therefore, we have

$$\int e^{at} \sin(bt) \, dt = \frac{1}{a}e^{at} \sin(bt) - \frac{b}{a} \left(\frac{1}{a}e^{at} \cos(bt) + \frac{b}{a} \int e^{at} \sin(bt) \right),$$

that is,

$$\int e^{at} \sin(bt) \, dt = \frac{e^{at} \sin(bt)}{a} - \frac{be^{at} \cos(bt)}{a^2} - \frac{b^2}{a^2} \int e^{at} \sin(bt).$$

Note the presence of $\int e^{at} \sin(bt) \, dt$ on both sides of the above expression. We can solve algebraically for this integral in the expression, thus obtaining

$$\int e^{at} \sin(bt) \, dt = -\frac{e^{ax}(b\cos(bx) - a\sin(bx))}{a^2 + b^2}.$$

Example 0.10

Use integration by parts to find a formula to compute the integral

$$I_n = \int_0^{\pi/2} \cos^n x \, dx,$$

where n is an non-negative integer. Use this formula to compute I_6 and I_7.

Solution: For $n = 0$ and $n = 1$, we get

$$I_0 = \frac{\pi}{2} \qquad \text{and} \qquad I_1 = 0.$$

Now suppose n is an integer greater than 1, and let us use integration by parts to find a formula to compute I_n. We begin by writing the integrand $\cos^n x$ as $\cos^{n-1} x \cos x$, and then we set

$$u = \cos^{n-1} x \quad \Rightarrow \quad du = -(n-1)\sin x \cos^{n-2} x \, dx$$
$$dv = \cos x \, dx \quad \Rightarrow \quad v = \sin x.$$

With this setting, formula (0.2) yields

$$I_n = \int_0^{\pi/2} \cos^n x \, dx = \cos^{n-1} x \sin x \, \Big|_0^{\pi/2} + \int_0^{\pi/2} (n-1)\cos^{n-2} x \sin^2 x \, dx,$$

where $\cos^{n-1} x \sin x \, \Big|_0^{\pi/2} = 0$. In the integral on the right hand side of the above equation, we can write $\sin^2 x = 1 - \cos^2 x$, and thus we have

$$\begin{aligned} I_n &= (n-1) \int_0^{\pi/2} \cos^{n-2} x (1 - \cos^2 x) \, dx \\ &= (n-1)\left[\int_0^{\pi/2} \cos^{n-2} x \, dx - \int_0^{\pi/2} \cos^n x \, dx \right]. \end{aligned}$$

We note that the last two integrals are in fact I_{n-2} and I_n, so

$$I_n = (n-1)(I_{n-2} - I_n) \quad \Rightarrow \quad I_n + (n-1)I_n = (n-1)I_{n-2},$$

and therefore the integral I_n can be written as

$$I_n = \frac{n-1}{n} I_{n-2}.$$

This is the formula that we were looking for. Let's now use it to compute I_6 and I_7. We have $I_6 = \frac{5}{6} I_4$, so the computation of I_6 involves the computation of I_4, but there is no problem with this, since we can use the formula repeatedly to compute I_n for decreasing values of n:

$$\begin{aligned} I_6 &= \frac{5}{6} I_4 \\ &= \frac{5}{6}\left(\frac{3}{4} I_2\right) = \frac{15}{24} I_2 \\ &= \frac{5}{8}\left(\frac{1}{2} I_0\right) = \frac{5\pi}{32}. \end{aligned}$$

A similar computation will show that

$$I_7 = \frac{16}{35} I_1,$$

but $I_1 = 0$, so $I_7 = 0$.

Sometimes we can evaluate an integral through a trigonometric change of variable when the integrand contains one of the expressions

$$\sqrt{a^2 - x^2} \qquad \sqrt{a^2 + x^2} \qquad \sqrt{x^2 - a^2}.$$

The following array indicates the type of substitution that can be done for each of these expressions:

$$\sqrt{a^2 - x^2},\ a > 0 \qquad\qquad x = a \sin \theta$$
$$\sqrt{a^2 + x^2},\ a > 0 \qquad\qquad x = a \tan \theta$$
$$\sqrt{x^2 - a^2},\ a > 0 \qquad\qquad x = a \sec \theta$$

Example 0.11

Evaluate

$$\int \frac{\sqrt{4 - x^2}}{x^2}\, dx.$$

Solution: Let $x = 2 \sin \theta$, so $dx = 2 \cos \theta\, d\theta$. Then

$$\sqrt{4 - x^2} = \sqrt{4 - 4 \sin^2 \theta} = 2\sqrt{\cos^2 \theta} = 2 \cos \theta.$$

Therefore,

$$
\begin{aligned}
\int \frac{\sqrt{4 - x^2}}{x^2}\, dx &= \int \frac{2 \cos \theta}{4 \sin^2 \theta} (2 \cos \theta\, d\theta) \\
&= \int \cot^2 \theta\, d\theta \\
&= \int (\csc^2 \theta - 1)\, d\theta \\
&= -\cot \theta - \theta + C
\end{aligned}
$$

Since $x = 2 \sin \theta$, $\theta = \arcsin \frac{1}{2} x$. We can determine what $\cot \theta$ is in terms of x by looking at the triangle below, which is constructed based on the fact that $\sin \theta = \frac{x}{2}$.

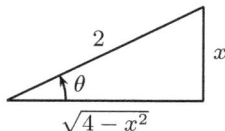

We see then that $\cot \theta = -\frac{\sqrt{9 - x^2}}{x}$, and

$$\int \frac{\sqrt{4 - x^2}}{x^2}\, dx = -\frac{\sqrt{9 - x^2}}{x} - \arcsin \frac{x}{2}.$$

Sometimes it is possible to integrate rational functions using their partial fraction decomposition. For example, the integral

$$\int \frac{7x-1}{x^2-x-6}\,dx$$

can be evaluated easily if we see that

$$\frac{7x-1}{x^2-x-6} = \frac{3}{x+2} + \frac{4}{x-3},$$

and therefore

$$\int \frac{7x-1}{x^2-x-6}\,dx = 3\ln|x+2| + 4\ln|x-3| + C.$$

To find the partial fraction decomposition of a rational function, we have to distinguish several cases. Assume $R(x) = \frac{p(x)}{q(x)}$ is a rational function that is a proper fraction, that is, the degree of the polynomial p is less than that of polynomial q.

Case 1   All factors of q are linear, and none of them is repeated. If $q(x)$ is factored as $q(x) = (a_1x+b_1)\cdots(a_nx+b_n)$, then

$$R(x) = \frac{A_1}{a_1x+b_1} + \cdots + \frac{A_n}{a_nx+b_n}$$

where $A_1, \ldots A_n$ are constants to be determined.

Case 2   All factors of q are linear, and some of them are repeated. Then, if $(ax+b)$ is repeated m times in the factorization of q, then to this factor corresponds the partial fraction decomposition

$$\frac{A_1}{ax+b} + \frac{A_2}{(ax+b)^2} + \cdots + \frac{A_m}{(ax+b)^m}.$$

Case 3   Some of the factors of q are quadratic, and none of them is repeated. Then to the quadratic factor (ax^2+bx+c) in the denominator corresponds the fraction

$$\frac{Ax+B}{ax^2+bx+c}.$$

Case 4   Some of the factors of q are quadratic, and some of them are repeated. If the factor (ax^2+bx+c) is repeated m times in the factorization of q, then to this factor corresponds the decomposition

$$\frac{A_1x+B_1}{ax^2+bx+c} + \frac{A_2x+B_2}{(ax^2+bx+c)^2} + \cdots + \frac{A_mx+B_m}{(ax^2+bx+c)^m}.$$

Example 0.12

Find the partial fraction decomposition of the following rational expressions:

(1) $\dfrac{7x-1}{x^2-x-6}$;

(2) $\dfrac{x^2-5x+2}{x(x-2)^2}$;

(3) $\dfrac{x^2-x-5}{(x-1)(x^2+2x+2)}$.

Verify the results using the *Mathematica* function **Apart**.

Solution: (1) The denominator can be factored as $x^2-x-6=(x+2)(x-3)$, so the partial fraction decomposition of this fraction is of the form

$$\frac{7x-1}{x^2-x-6}=\frac{A}{x+2}+\frac{B}{x-3}.$$

Adding the fractions in the right-hand side of the above expression yields the numerator $A(x-3)+B(x+2)$, so it must be

$$A(x-3)+B(x+2)=7x-1.$$

Since this equality must hold for all values of x, we can choose any value of x in order to solve for A and B. If $x=-2$, we get $-5A+0=-15$, so $A=3$. Choosing $x=3$ gives $0+5B=20$, so $B=4$, and the partial fraction decomposition is

$$\frac{7x-1}{x^2-x-6}=\frac{3}{x+2}+\frac{4}{x-3}.$$

(2) The factorization of the denominator contains $(x-2)^2$, so the factor $(x-2)$ is repeated twice. Hence, case 2 applies, and the partial fraction decomposition is of the form

$$\frac{x^2-5x+2}{x(x-2)^2}=\frac{A}{x}+\frac{B}{x-2}+\frac{C}{(x-2)^2}.$$

Adding the fractions in the right-hand side yields the numerator

$$(A+B)x^2+(-4A-2B+C)x+4A=x^2-5x+2.$$

Since the above equality must hold for all values of x, the coefficients in the right-hand side must match the coefficients in the left-hand side. More exactly, we must have

$$\begin{aligned}
A+B &= 1\\
-4A-2B+C &= -5\\
4A &= 2.
\end{aligned}$$

Therefore, $A = 1/2$, $B = 1/2$ and $C = -2$, and the partial fraction decomposition is

$$\frac{x^2 - 5x + 2}{x(x-2)^2} = \frac{1}{2x} + \frac{1}{2(x-2)} - \frac{2}{(x-2)^2}.$$

(3) The denominator contains a linear factor $(x-1)$ and a quadratic factor $(x^2 + 2x + 2)$ that cannot be factored further into linear factors, so case 3 applies, and the partial fraction decomposition is of the form

$$\frac{x^2 - x - 5}{(x-1)(x^2 + 2x + 2)} = \frac{A}{x-1} + \frac{Bx + C}{x^2 + 2x + 2}.$$

In this case we must have

$$(Bx + C)(x - 1) + A(x^2 + 2x + 2) = x^2 - x - 5,$$

We can compute A by substituting $x = 1$, which gives us $0 + 5A = -5$, so $A = -1$. Substituting $A = -1$ and multiplying out in the left-hand side gives

$$(B - 1)x^2 + (-B + C - 2)x - C - 2 = x^2 - x - 5,$$

so $(B - 1) = 1$ and $(-C - 2) = -5$, and then $B = 2$ and $C = 3$. The partial fraction decomposition is therefore

$$-\frac{1}{x-1} + \frac{2x + 3}{x^2 + 2x + 2}.$$

We now find the all the above decompositions using the *Mathematica* function **Apart**:

In[1]:= **Apart**$\left[\frac{7\,x-1}{x^2-x-6}\right]$

Out[1]:= $\frac{4}{-3+x} + \frac{3}{2+x}$

In[2]:= **Apart**$\left[\frac{x^2-5\,x+2}{x\,(x-2)^2}\right]$

Out[2]:= $-\frac{2}{(-2+x)^2} + \frac{1}{2\,(-2+x)} + \frac{1}{2\,x}$

In[3]:= **Apart**$\left[\frac{x^2-x-5}{(x-1)\,(x^2+2\,x+2)}\right]$

Out[3]:= $-\frac{1}{-1+x} + \frac{3+2\,x}{2+2\,x+x^2}$

An important technique for solving differential equations involves the computation of a special improper integral that yields the so-called **Laplace transformation** of a function. Before showing you what this integral is, let us recall the definition of an improper integral. If f is a continuous function for all values in the interval $[a, \infty)$, then we define

$$\int_a^\infty f(x) \, dx = \lim_{b \to +\infty} \int_a^b f(x) \, dx,$$

We say that the above integral converges or that it diverges according to whether the limit in the right-hand side exists or not, respectively. Similarly, if f is continuous in $(-\infty, b]$, then we define

$$\int_{-\infty}^b f(x) \, dx = \lim_{a \to -\infty} \int_a^b f(x) \, dx.$$

If the function f is continuous in the entire $\mathbb{R}$ and c is any real number, then we define

$$\int_{-\infty}^\infty f(x) \, dx = \int_{-\infty}^c f(x) \, dx + \int_c^\infty f(x) \, dx.$$

The convergence or divergence of the above integral is independent of what value of c we choose.

Example 0.13

Evaluate the following integral.

$$\int_{-\infty}^\infty \frac{1}{x^2 + 1} \, dx.$$

Solution: We have

$$
\begin{aligned}
\int_{-\infty}^\infty \frac{1}{x^2+1} \, dx &= \int_{-\infty}^0 \frac{1}{x^2+1} \, dx + \int_0^\infty \frac{1}{x^2+1} \, dx \\
&= \lim_{a \to -\infty} [\arctan x]_a^0 + \lim_{b \to \infty} [\arctan x]_a^0 \\
&= 0 - \lim_{a \to -\infty} (\arctan a) + \lim_{b \to \infty} (\arctan b) - 0 \\
&= -\left(-\frac{\pi}{2}\right) + \frac{\pi}{2} \\
&= \pi.
\end{aligned}
$$

There are some important improper integrals that cannot be written using elementary functions and that arise in many applications. One of these integrals that will be useful to us is the **Gamma function**, defined as

$$\Gamma(x) = \int_0^\infty t^{x-1} e^{-t} \, dt \tag{0.3}$$

The gamma function is a generalization of the factorial function, and for $x = n \in \{1, 2, 3, \ldots\}$, we have

$$\Gamma(n) = (n-1)!.$$

To show this, we apply integration by parts to **(0.3)**, defining $u = t^{n-1}$ and $dv = e^{-t}\, dt$, so $du = (n-1)t^{n-2}\, dt$ and $v = -e^{-t}$. Therefore,

$$
\begin{aligned}
\Gamma(n) &= \int_0^\infty t^{n-1} e^{-t}\, dt \\
&= -t^{n-1} e^{-t}\Big|_0^\infty + \int_0^\infty (n-1) t^{n-2} e^{-t}\, dt \\
&= 0 + (n-1) \int_0^\infty t^{n-2} e^{-t}\, dt \\
&= (n-1)\Gamma(n-1).
\end{aligned}
$$

We then apply the same technique repeatedly to obtain

$$
\begin{aligned}
\Gamma(n) &= (n-1)\Gamma(n-1) \\
&= (n-1)(n-2)\Gamma(n-2) \\
&= (n-1)(n-2)(n-3)\cdots\Gamma(1)
\end{aligned}
$$

where it is immediate that $\Gamma(1) = 1$. Hence, $\Gamma(n) = (n-1)!$ as desired.

0.4.1   The Laplace Transform

Definition 0.14
Let f be a function defined on the interval $[0, \infty)$. The **Laplace transform** of f, denoted $\mathcal{L}[f]$, is a function F given by

$$F(s) = \mathcal{L}[f(x)] = \int_0^\infty f(x) e^{-sx}\, dx$$

provided this integral exists.

In the following examples, we will obtain the Laplace transform of several elementary functions.

Example 0.15

Compute $\mathcal{L}[f]$ for f given by $f(x) = 1$, and state for which values of s this Laplace transform exists.

Solution: By definition,

$$\mathcal{L}[1] = \int_0^\infty 1 e^{-sx} = \int_0^\infty e^{-sx}\, dx. \tag{0.4}$$

If $s = 0$, the improper integral **(0.4)** diverges, since

$$\int_0^\infty dx = \lim_{b \to +\infty} b = +\infty.$$

Now suppose $s \neq 0$. Then

$$
\begin{aligned}
\int_0^\infty e^{-sx}\,dx &= \left.\frac{e^{-sx}}{-s}\right|_0^\infty \\
&= \lim_{b \to +\infty} \frac{-1}{s}(e^{-sb} - 1).
\end{aligned}
$$

This limit is $1/s$ provided $s > 0$, and it diverges to $+\infty$ otherwise. Hence,

$$
\mathcal{L}\,[1] = \frac{1}{s}, \qquad s > 0.
$$

Example 0.16

Find the Laplace transform of f given by $f(x) = e^{ax}$.

Solution:

$$
\mathcal{L}\,[e^{ax}] = \int_0^\infty e^{ax}e^{-sx}\,dx = \int_0^\infty e^{-(s-a)x}\,dx = \left.\frac{-1}{s-a}e^{-(s-a)x}\right|_0^\infty = \frac{1}{s-a}
$$

whenever $s > a$, and for $s \leq a$ the transform $\mathcal{L}\,[e^{ax}]$ does not exist.

Example 0.17

Compute the Laplace transform of f given by $f(x) = x^n$, with $n \in \{1, 2, 3, \ldots\}$.

Solution: We have

$$
\mathcal{L}\,[x^n] = \int_0^\infty x^n e^{-sx}\,dx.
$$

We use integration by parts to calculate this integral: if we define $u = x^n$ and $dv = e^{-sx}\,dx$, then

$$
du = nx^{n-1} \qquad \text{and} \qquad v = -\frac{1}{s}e^{-sx},
$$

so integration by parts gives

$$
\int_0^\infty x^n e^{-sx}\,dx = \left.-\frac{x^n}{s}e^{-sx}\right|_0^\infty + \frac{n}{s}\int_0^\infty x^{n-1}e^{-sx}\,dx = \frac{n}{s}\mathcal{L}\,[x^{n-1}]
$$

provided $s > 0$, for in this case $\left.-\frac{x^n}{s}e^{-sx}\right|_0^\infty = 0$. By induction, we obtain

$$
\mathcal{L}\,[x^n] = \frac{n}{s}\mathcal{L}\,[x^{n-1}] = \frac{n(n-1)}{s^2}\mathcal{L}\,[x^{n-2}] = \cdots = \frac{n!}{s^n}\mathcal{L}\,[1] = \frac{n!}{s^{n+1}}
$$

for $s > 0$.

The Laplace transform of $f(x) = x^p$, where p is any real number > -1, can be written in terms of the Gamma function. If we define $t = sx$, then $dt = sdx$, and

$$\mathcal{L}\left[x^p\right] = \int_0^\infty x^p e^{-sx} \, dx \tag{0.5}$$

$$= \int_0^\infty \left(\frac{t}{s}\right)^p e^{-t} \left(\frac{1}{s}\right) dt \tag{0.6}$$

$$= \frac{1}{s^{p+1}} \int_0^\infty t^p e^{-t} \, dt \tag{0.7}$$

$$= \frac{1}{s^{p+1}} \Gamma(p+1). \tag{0.8}$$

Example 0.18

Find the Laplace transform of f given by $f(x) = \sin ax$.

Solution: We have

$$\mathcal{L}\left[\sin ax\right] = \int_0^\infty (\sin ax) e^{-sx} \, dx.$$

Again, to evaluate this integral, we use integration by parts. Let $u = \sin ax$ and $dv = e^{-sx} \, dx$. Then

$$\int_0^\infty (\sin ax) e^{-sx} \, dx = \left. \frac{-1}{s} (\sin ax) e^{-sx} \right|_0^\infty + \frac{a}{s} \int_0^\infty (\cos ax) e^{-sx} \, dx$$

$$= \frac{a}{s} \int_0^\infty (\cos ax) e^{-sx} \, dx \qquad \text{provided } s > 0.$$

Using integration by parts again to compute this last integral, we find that

$$\mathcal{L}\left[\sin ax\right] = \frac{a}{s^2} - \frac{a^2}{s^2} \mathcal{L}\left[\sin ax\right],$$

so

$$\mathcal{L}\left[\sin ax\right] = \frac{\frac{a}{s^2}}{1 + \frac{a^2}{s^2}} = \frac{a}{s^2 + a^2}, \qquad s > 0.$$

If f is a function and $F(s) = \mathcal{L}\left[f(x)\right]$, then differentiating F with respect to s gives

$$\frac{d}{ds} F(s) = \frac{d}{ds} \int_0^\infty f(x) e^{-sx} \, dx$$

$$= \int_0^\infty \frac{d}{ds} f(x) e^{-sx} \, dx$$

$$= -\int_0^\infty x f(x) e^{-sx} \, dx$$

$$= -\mathcal{L}\left[x f(x)\right].$$

This result can be generalized to the n-th derivative of F by means of induction. This will give us

$$\frac{d^n}{ds^n} F(s) = (-1)^n \mathcal{L}\left[x^n f(x)\right]. \tag{0.9}$$

Example 0.19

Calculate $\mathcal{L}\left[x \sin ax\right]$.

Solution: We have seen already that $\mathcal{L}\left[\sin ax\right] = \frac{a}{s^2+a^2}$, and by the formula (0.9), we have

$$\mathcal{L}\left[x \sin ax\right] = -\frac{d}{ds}\left(\frac{a}{s^2+a^2}\right) = \frac{2as}{(s^2+a^2)^2}.$$

Another useful result is the **translation property** of the Laplace transform, which we now state and prove.

Theorem 0.20
If $\mathcal{L}\left[f(x)\right] = F(s)$ for $s > s_0$, then $\mathcal{L}\left[e^{cx}f(x)\right] = F(s-c)$ for $s > s_0 + a$.

Proof: By definition,

$$\begin{aligned}
\mathcal{L}\left[e^{cx}f(x)\right] &= \int_0^\infty e^{cx} f(x) e^{-sx}\, dx \\
&= \int_0^\infty f(x) e^{-(s-c)x}\, dx,
\end{aligned}$$

but by hypothesis, the last integral is $F(s-c)$ for $s-c > s_0$, that is, $\mathcal{L}\left[e^{cx}f(x)\right] = F(s-c)$ for $s > s_0 + a$ as desired. ∎

Example 0.21

Evaluate $\mathcal{L}\left[e^{cx}\sin ax\right]$.

Solution: By the translation property (theorem 0.20), we have

$$\mathcal{L}\left[e^{cx}\sin ax\right] = F(s-c),$$

where $F(s) = \mathcal{L}\left[\sin ax\right] = \frac{a}{s^2+a^2}$, so

$$\mathcal{L}\left[e^{cx}\sin ax\right] = \frac{a}{(s-c)^2+a^2}.$$

Using the unit step function u, we can formulate a **second translation property** of the Laplace transform.

Theorem 0.22

Let $\mathcal{L}[f(x)]$ exist for $s > a \geq 0$ and $c > 0$. Then

$$\mathcal{L}[u(x-c)f(x-c)] = e^{-cs}\mathcal{L}[f(x)].$$

Proof:

$$\mathcal{L}[u(x-c)f(x-c)] = \int_0^\infty u(x-c)f(x-c)e^{-sx}\,dx = \int_c^\infty f(x-c)e^{-sx}\,dx,$$

and defining $\xi = x - c$, we have

$$
\begin{aligned}
\mathcal{L}[u(x-c)f(x-c)] &= \int_c^\infty f(x-c)e^{-sx}\,dx \\
&= \int_c^\infty f(\xi)e^{-s(\xi+c)}\,d\xi \\
&= e^{-cs}\int_0^\infty f(\xi)e^{-s\xi}\,d\xi = e^{-cs}\mathcal{L}[f(x)].
\end{aligned}
$$

∎

There are some properties of the Laplace transform that are of particular interest to us, since they will allow us to use Laplace transformation to solve differential equations in later chapters. The example below introduces one of these properties.

Example 0.23

If f is a differentiable function and $\mathcal{L}[f(x)] = F(s)$ for $s > s_0$, compute $\mathcal{L}[f'(x)]$.

Solution: By definition,

$$\mathcal{L}[f'(x)] = \int_0^\infty f'(x)e^{-sx}\,dx,$$

and using integration by parts,

$$
\begin{aligned}
\int_0^\infty f'(x)e^{-sx}\,dx &= f(x)e^{-sx}\Big|_0^\infty + \int_0^\infty sf(x)e^{-sx}\,dx \\
&= -f(0) + s\mathcal{L}[f(x)].
\end{aligned}
$$

Hence,

$$\mathcal{L}[f'(x)] = s\mathcal{L}[f(x)] - f(0).$$

0.4.2   Inverse of the Laplace Transform

One of the main characteristics of the Laplace transform is that it is invertible when it is applied to continuous functions. This is a consequence of the theorem that we now state.

Theorem 0.24
Let f and g be continuous functions, and suppose that $F(s) = \mathcal{L}[f(x)]$ and $G(s) = \mathcal{L}[g(x)]$ exist and $F(s) = G(s)$. Then $f(s) = g(s)$.

The theorem above implies that a function f can be determined from its Laplace transform. The operation performed to go from $F(s)$ to $f(x)$ is called the inverse Laplace transform, and it is denoted and given by

$$\mathcal{L}^{-1}[F(s)] = f(x) \qquad \text{if and only if} \qquad \mathcal{L}[f(x)] = F(s).$$

Just as the Laplace transform, the inverse Laplace transform is a linear operator, meaning

$$\mathcal{L}^{-1}[c_1 F(s) + c_2 G(s)] = c_1 \mathcal{L}^{-1}[F(s)] + c_2 \mathcal{L}^{-1}[G(s)].$$

Let us now see an example in which we compute the inverse Laplace transform of some functions.

Example 0.25
Find $\mathcal{L}^{-1}[F(s)]$ if

(1) $F(s) = \dfrac{2}{s^2 + 1} + \dfrac{3}{s-1}$.

(2) $F(s) = \dfrac{3s^2 + 2}{(s-1)(s^2+1)}$.

Solution: (1) We have

$$\mathcal{L}^{-1}\left[\frac{2}{s^2+1} + \frac{3}{s-1}\right] = 2\mathcal{L}^{-1}\left[\frac{1}{s^2+1}\right] + 3\mathcal{L}^{-1}\left[\frac{1}{s-1}\right],$$

and we know that

$$\mathcal{L}[\sin x] = \frac{1}{s^2+1}$$

$$\mathcal{L}[e^x] = \frac{1}{s-1},$$

so

$$\mathcal{L}^{-1}\left[\frac{2}{s^2+1} + \frac{3}{s-1}\right] = 2\sin x + 3e^x.$$

(2) We do not have a formula that directly gives us the inverse Laplace transform of $F(s) = \dfrac{3s^2 + 2}{(s-1)(s^2+1)}$. Here is where partial fraction decomposition

becomes handy, for we can decompose $F(s)$ into the sum of simpler fractions whose inverse Laplace transform is known. We find that

$$\frac{3s^2 + 2}{(s-1)(s^2+1)} = \frac{s+1}{2(s^2+1)} + \frac{5}{2(s-1)}$$

$$= \frac{s}{2(s^2+1)} + \frac{1}{2(s^2+1)} + \frac{5}{2(s-1)},$$

so

$$\mathcal{L}^{-1}[F(s)] = \frac{1}{2}\mathcal{L}^{-1}\left[\frac{s}{(s^2+1)}\right] + \frac{1}{2}\mathcal{L}^{-1}\left[\frac{1}{(s^2+1)}\right] + \frac{5}{2}\mathcal{L}^{-1}\left[\frac{1}{(s-1)}\right]$$

$$= \frac{\cos(x)}{2} + \frac{\sin(x)}{2} + \frac{5e^x}{2}.$$

Often we have to compute the inverse Laplace transform of the product of the Laplace transforms of two functions. In this case, we cannot always resort to partial fraction decomposition. In such cases, we will need an additional result that involves an operation called **convolution**.

Definition 0.26

The **convolution** of two functions f and g is denoted and given by

$$(f * g)(x) = \int_0^x f(\xi)g(x-\xi)\, d\xi,$$

provided this integral exists for $x > 0$.

Example 0.27

Given

$$f(x) = \sin x \qquad \text{and} \qquad g(x) = x,$$

compute $(f * g)(x)$.

Solution: By definition,

$$(\sin x * x)(x) = \int_0^x \sin\xi(x-\xi)\, d\xi$$

$$= x\int_0^x \sin\xi\, d\xi - \int_0^x \xi\sin\xi\, d\xi$$

$$= x - x\cos - \sin x + x\cos x$$

$$= x - \sin x.$$

We can now state the following result, called the **convolution theorem**.

Theorem 0.28
Let $F(s) = \mathcal{L}\left[f(x)\right]$ and $G(s) = \mathcal{L}\left[g(x)\right]$. Then

$$F(s)G(s) = \mathcal{L}\left[(f * g)(x)\right].$$

$f(x)$	$F(s) = \mathcal{L}\left[L\right]\left[f(x)\right]$
x^n, $n \in \{1, 2, 3, \ldots\}$	$\dfrac{n!}{s^{n+1}}$
x^r, $r > -1$	$\dfrac{\Gamma(r+1)}{s^{r+1}}$
e^{ax}	$\dfrac{1}{s-a}$
$\sin(ax)$	$\dfrac{a}{s^2 + a^2}$
$\cos(ax)$	$\dfrac{s}{s^2 + a^2}$
$\sin(ax + b)$	$\dfrac{s\sin(b) + a\cos(b)}{s^2 + a^2}$
$\cos(ax + b)$	$\dfrac{s\cos(b) - a\sin(b)}{s^2 + a^2}$
$\sinh(ax)$	$\dfrac{a}{s^2 - a^2}$
$\cosh(atx$	$\dfrac{s}{s^2 - a^2}$
$e^{ax}f(x)$	$F(s - c)$
$u(x - c)$	$\dfrac{e^{-cs}}{s}$
$u(x - c)f(x - c)$	$e^{-cs}F(s)$
$t^n f(x)$, $n \in \{1, 2, 3, \ldots\}$	$(-1)^n F^{(n)}(s)$
$f'(x)$	$sF(s) - f(0)$
$f^{(n)}(x)$	$s^n F(s) - s^{n-1}f(0) - s^{n-2}f'(0) - \cdots - sf^{(n-2)}(0) - f^{(n-1)}(0)$
$f(x) + g(x)$	$F(s) + G(s)$
$(f * g)(x) = \displaystyle\int_0^x f(x - \xi)g(\xi)\, d\xi$	$F(s)G(s)$

Refreshing the material contained in this chapter is absolutely essential to get through the rest of the book and understand the course very well.

Exercises

(1) Given

$$T = \frac{4}{\omega_0} \int_0^1 \frac{1}{\sqrt{1 - u^2 + z - zu^5}}\, dz,$$

find

$$\lim_{z \to 0} T.$$

(2) A function f is **periodic** if there exists a T such that

$$f(t + T) = f(t), \qquad \text{for all } t \text{ in the domain of } f.$$

Are the following functions periodic?

    (a) $f(t) = \sin t$.

    (b) $g(t) = \cos t$.

(3) Find the domain and range of the following functions.

    (a) $f(x) = -\sqrt{x^2 - 1}$.

    (b) $g(x) = \cos\left(\frac{Pix}{4}\right)$

    (c) $h(x) = \frac{2}{x-1}$

    (d) $f(x) = \begin{cases} 2x + 1, & x < 0 \\ 2x + 2, & x \geq 0 \end{cases}$

    (e) $f(x) = \begin{cases} |x| + 1, & x < 1 \\ -x + 1, & x \geq 1 \end{cases}$

(4) Find the x-values (if any) at which f is not continuous.

    (a) $f(x) = \frac{x-1}{x^2 - 3x + 2}$

    (b) $f(x) = \frac{|x-1|}{x-1}$

    (c) $f(x) = \begin{cases} -2x + 3, & x < 1 \\ x^2, & x \geq 1 \end{cases}$

(5) Find the derivative of the following functions.

    (a) $y(t) = \sqrt{t^3 - 4t^2 + 1}$

    (b) $y(x) = \sin(\tan 2x)$

    (c) $g(\theta) = \cos^2(4\theta)$

(6) Find the polynomial $p(x) = a_0 + a_1 x$ whose value and slope agree with the value and slope of $f(x) = \cos x$ at $x = 0$.

(7) Let f be a function satisfying $f(0) = f'(0) = 1$. Prove that if $f(a + b) = f(a)f(b)$ for all a and b, then f is differentiable and

$$f'(x) = f(x) \qquad \text{for all } x.$$

Can you predict one function that satisfies this property?

(8) Evaluate the following integrals.

(a) $\displaystyle\int (5\cos x + 4\sin x)\, dx$

(b) $\displaystyle\int \frac{\sin x}{1 - \sin^2 x}\, dx$

(c) $\displaystyle\int_0^1 2\sin 3x\, dx$

(d) $\displaystyle\int x^2 \cos x\, dx$

(e) $\displaystyle\int_{-\pi}^{\pi} \sin^2 x\, dx$

(f) $\displaystyle\int_0^{\pi/2} \frac{\cos x}{1 + \sin x}\, dx$

(g) $\displaystyle\int_0^1 \sin^{378} x\, dx$

(h) $\displaystyle\int \frac{\sqrt{x^2 - 9}}{x}\, dx$ (Useful integral to solve the DE $x\dfrac{dy}{dx} = \sqrt{x^2 - 9}$)

(i) $\displaystyle\int \frac{dx}{\sqrt{x^2 + 4}}\, dx$ (Useful integral to solve the DE $\sqrt{x^2 + 4}\,\dfrac{dy}{dx} = 1$)

(9) Evaluate the following integrals using partial fraction decomposition.

(a) $\displaystyle\int \frac{1}{x^2 - 16}\, dx$

(b) $\displaystyle\int \frac{x + 1}{x^2 + 11x + 18}\, dx$

(c) $\displaystyle\int \frac{x^2 + x + 3}{x^4 + 6x^2 + 9}\, dx$

(d) $\displaystyle\int \frac{1}{x(a + bx)}\, dx$

(10) Use the definition and properties of the Laplace transform to compute the Laplace transform of the following functions.

(a) $f(x) = 3$

(b) $f(x) = 2x$

(c) $f(x) = e^{2x}$

(d) $f(x) = te^x$

(e) $f(x) = \sin 2x$

(f) $f(x) = \cos 3x$

(g) $f(x) = e^{-ax}$

(h) $f(x) = \sin(4x)$

(i) $f(x) = \cosh x$

(j) $f(x) = t^2 + t$

(k) $f(x) = e^{3x}\cos(x)$

(l) $u_3(x)(x-3)$

(m) $u_7(\pi/2)\sin x$

(11) Use the table of Laplace transforms at the end of this chapter to compute the inverse Laplace transform of the following functions.

(a) $\dfrac{s}{s^2+4}$

(b) $\dfrac{5}{s^2+25}$

(c) $\dfrac{2}{s^2+16}$

(d) $\dfrac{2s}{s(s^2-3s+2)}$

(e) $\dfrac{10s^2-5}{(s-2)(s+1)(s+3)}$

(f) $\dfrac{e^{2s}}{s-3}$

(12) Prove that
$$\int_0^1 \frac{x^4(1-x)^4}{1+x^2}\,dx = \frac{22}{7} - \pi.$$
(Composed by the committee of the Putnam Prize competition.)

(13) Verify the following.

(a) $\displaystyle\int \frac{1}{(x^2+a^2)^{3/2}}\,dx = \frac{x}{a^2\sqrt{x^2+a^2}}$

(b) $\displaystyle\int \arctan u\,du = u\arctan u - \ln(1+u^2) + C$

Chapter 1

Preliminaries

In this chapter we begin to introduce the fundamental concepts of the theory of differential equations. Naturally, the first definition of this chapter is that of a differential equation. Next, we will study the concept of a solution to a differential equation, where we will spend some time looking at examples that will clarify the details involved in determining whether we have a solution to a differential equation or not. Section 1.1.3 studies families of solutions to differential equations. Section 1.1.4 contains the last topic of this chapter that is directly related to differential equations, namely the study of direction fields, which constitute a geometrical technique to explore the nature of the solutions to a differential equation. All sections that follow will be devoted to an introduction to linear algebra, which is merely a presentation of the mathematical language that will be useful in later chapters, as well as some results that are necessary in our study of differential equations.

1.1 Basic Definitions

1.1.1 What Is a Differential Equation?

When modeling a real phenomenon it is usually difficult, or impossible, to find an expression (or model) that relates the quantities involved in the phenomenon in such a way that only algebraic operations and elementary functions appear in the expression. In particular, we often find that considering the **rate of change** of the quantities of the phenomenon is fundamental to our description of it, and thus it is likely that an equation that satisfactorily models this phenomenon includes the **derivatives** of one or more of the quantities involved. An example of this is Newton's second law,

$$m\frac{d^2u(t)}{dt^2} = F\left(t, u(t), \frac{du(t)}{dt}\right), \tag{1.1}$$

which relates the position $u(t)$ of a particle acted on by a force F, which in turn may be a function of time t, the position $u(t)$ and the velocity $\frac{du(t)}{dt}$. Thus, in

31

order to determine the position of a particle it is necessary to find a **solution** for **(1.1)**, that is, to find a function $u(t)$ that satisfies the equation **(1.1)**.

An equation like **(1.1)** that includes the derivative of an unknown function f and one or more of the derivatives of f is called a **differential equation**. For example, the following equation includes the first derivative of a function y of t, and so is a differential equation:

$$y'(t) - y(t) - t^2 = 0.$$

Before introducing more examples of differential equations, let us agree on a notational convention:

Notation: Usually the context clarifies which symbol is used to represent the independent variable and which symbol is used to represent the function of that variable. When this is so, one can simply write y instead of $y(t)$, for it will be clear from the context that y is a function of t. This can be done, of course, when symbols other than t and y are used. For example, if it is clear that y is a function of x, then the differential equation

$$y''(x) + 3y'(x) - y(x) = \cos(x)$$

may be written as

$$y'' + 3y' - y = \cos(x).$$

Here are some other examples of differential equations:

$$y' - y^2 - t = 0, \tag{1.2}$$
$$s'' - 2t^2 s' - 7 = 0, \tag{1.3}$$
$$\frac{d^2y}{dt^2} + \omega_0 y = F\sin(\omega t + \beta), \tag{1.4}$$
$$y''' = y'' + y' + y, \tag{1.5}$$
$$\frac{dy}{dx} = \frac{1}{x^2 + y^2}, \tag{1.6}$$
$$y^{(5)} - y'' = e^{x^2} - \sin(xy). \tag{1.7}$$

Let us now introduce the definition of ordinary differential equation more formally.

Definition 1.1

Let $y = f(x)$ define y as a function of x in some interval (a,b). An **ordinary differential equation** for the function y is an equation involving the independent variable x, the function y and one or more of the derivatives of y. Thus, a differential equation is an equation of the form

$$F\left(x, y, y'', y''', \dots, y^{(n)}\right) = 0. \tag{1.8}$$

Note: The derivative of functions of more than one variable are partial derivatives, and equations involving this type of derivatives are called **partial differential equations** (PDE) to distinguish them from the **ordinary differential equations** (ODE), which include only derivatives of functions of a single variable, like in the examples above. Examples of partial differential equations are the heat equation and the wave equation:

$$\frac{\partial u}{\partial t} = \frac{\partial^2 u}{\partial x^2} \qquad \text{and} \qquad \frac{\partial^2 u}{\partial^2 t} = \frac{\partial^2 u}{\partial x^2}.$$

Here we will only study ordinary differential equations. Therefore, in our discussion, the terminology "differential equation" will always refer to an *ordinary* differential equation.

A general way of classifying differential equations is according to their order, which we now define.

Definition 1.2

The **order** of a differential equation is defined as the order of the highest derivative in the equation.

For example, **(1.2)** and **(1.6)** are first order differential equations (or differential equations of order one), **(1.3)** and **(1.4)** are second order differential equations, **(1.5)** is a differential equation of order three, and **(1.7)** is a differential equation of order five.

1.1.2   Solution to a Differential Equation

We will now look at the most important definition of this section, namely, that of the solution to a differential equation. The student should spend some time analyzing this definition until it is well understood. In particular, we ask the student to read the comments and remarks that follow the definition.

Definition 1.3

We say that a *function* φ is a **solution** to the differential equation

$$F\left(x, y, y', \ldots, y^{(n)}\right) = 0 \tag{1.9}$$

in an *interval* (a, b), if $\varphi(t)$ *satisfies* the equation for all t in (a, b), that is, if substituting $y(t)$ by $\varphi(t)$ in **(1.9)** yields a true statement for all values of t in (a, b). Usually, a solution φ to **(1.9)** is denoted by y, that is, by the same symbol used to represent the dependent variable in a differential equation.

Starting in chapter 2, we will begin studying techniques to find relations $y = f(x)$ (and sometimes implicit relations $f(x, y) = 0$) that define solutions to differential equations. However, the student should always keep in mind that a solution is more than such relations, as can be judged from the preceding definition, which tells us that the relation between y and x established by an expression $y = f(x)$ (or an implicit relation $f(x, y) = 0$) must be that of a *function* for it to be a candidate for a solution to our given differential equation. If the expression does define y as a function of x, we can then proceed to test if this function *satisfies* the equation on an *interval* in which we claim it is a solution. Taking into account this **interval of validity** is absolutely necessary, for it tells us where our solution is really a solution. As a matter of fact, sometimes we will not mention the interval of validity explicitly when giving a solution to a differential equation, but this is just because we expect the student to determine it.

Now let's look at some examples of solutions to differential equations. At this point, you don't need to know how these solutions were obtained, for that will be studied in subsequent chapters. For now, we will only focus on showing that a given solution is indeed a solution, all based on the definition 1.3. '

Example 1.4

Show that the differential equation

$$y' - y = x \tag{1.10}$$

has the function y, given by $y(t) = 2e^x - x - 1$, as a solution in the entire set of real numbers $\mathbb{R}$.

Solution: Since we see that the expression $y(t) = 2e^x - x - 1$ does define y as a function of x, we can proceed to verify that $y(x)$ satisfies the differential equation (**1.10**), that is, that we do have

$$y' - y = x.$$

To do this, we compute $y'(x) = 2e^x - 1$, so substituting into the differential equation gives

$$
\begin{aligned}
y' - y &= x \\
2e^x - 1 - (2e^x - x - 1) &= x \\
x &= x,
\end{aligned}
$$

so $y(t) = 2e^x - x - 1$ satisfies the differential equation for any t in $\mathbb{R}$. Hence, this function y is a solution to (**1.10**) with interval of validity $\mathbb{R}$.

Example 1.5

Show that the relation $y(x) = \sqrt{2x+1}$ defines a solution y to the differential equation

$$y' = \frac{1}{y},$$

in the interval $(-1/2, \infty)$.

Solution: Let us proceed by showing that the function y given by $y(x) = \sqrt{2x+1}$ satisfies the differential equation. Indeed,

$$y'(x) = \frac{2}{2\sqrt{2x+1}} = \frac{1}{\sqrt{2x+1}},$$

where the right-hand side is $1/y(x)$, so we do have $y' = 1/y$, which means that $y(x) = \sqrt{2x+1}$ satisfies the differential equation. Now see that $y(x)$ is not a real number for $x < -1/2$, and that $y(x)$ is zero when $x = -1/2$. Hence, for $1/y(x)$ to be defined and to be real, we must have $x \in (-1/2, \infty)$, which is the interval of validity given for the solution $y(x) = \sqrt{2x+1}$.

The following example illustrates how to use *Mathematica* to verify that a function is a solution to a differential equation.

Example 1.6

Use *Mathematica* to show that the function y given by $y(x) = (x+1)e^x$ is a solution to the differential equation

$$y'' - 2y' + y = 0.$$

Solution: We define the function $y(x) = (x+1)e^x$ in *Mathematica*:

In[1]:= `Clear[y]; y[x_] := (x + 1) e`^{x}

Out[1]:=

Then we evaluate

In[2]:= `Simplify[ (y')'[x] - 2 y'[x] + y[x] == 0]`

Out[2]:= `True`

This results shows that the equality $y''(x) - 2y'(x) + y(x) = 0$ is true for all values of x, so $y(x) = (x+1)e^x$ is a solution to the DE $y'' - 2y' + y = 0$.

1.1.3 Family of Solutions

Consider the differential equation

$$y' = x. \tag{1.11}$$

One can verify immediately that all the three functions y_1, y_2 and y_3 given by

$$y_1(x) = \frac{1}{2}x^2, \quad y_2(x) = \frac{1}{2}x^2 - 1, \quad y_3(x) = \frac{1}{2}x^2 + 7 \tag{1.12}$$

are solutions to (1.11). In fact, note that any function y of x of the form

$$y = \frac{1}{2}x^2 + c, \tag{1.13}$$

where c is an arbitrary constant, is a solution to (1.11), and that the solutions in (1.12) are obtained by choosing specific values for the constant c. We see then that the differential equation in (1.11) actually has infinitely many solutions, each of which is determined by substituting a specific value for the constant c in **(1.11)**.

Let us consider now the differential equation

$$y'' = x. \tag{1.14}$$

The reader can verify that the following are solutions of this differential equation:

$$y_1(x) = \frac{1}{6}x^3, \quad y_2(x) = \frac{1}{6}x^3 - x + 1, \quad y_3(x) = \frac{1}{6}x^3 + 3x - 2. \tag{1.15}$$

In this case, we have that any function y of x given by

$$y = \frac{1}{6}x^3 + c_1 x + c_2,$$

with c_1 and c_2 arbitrary constants, is a solution to **(1.14)**. Again, the solutions in **(1.15)** are obtained by choosing specific values for c_1 and c_2.

The examples above illustrate the fact that a differential equation may have infinitely many solutions, and that each of this solutions may be obtained from a certain expression containing some parameters, like c_1 and c_2 in the example above. This leads us to consider the concept of a **family of solutions**.

Definition 1.7

A function f depending on x and on the m parameters $c_1, \ldots, c_m$ is called an **m-parameter family of solutions** of the nth order differential equation

$$F\left(x, y, y', y'', \ldots, y^{(n)}\right) = 0 \tag{1.16}$$

if for each choice of a set of values for $c_1, \ldots c_m$ the function φ defined by $\varphi(x) = f(x, c_1, \ldots, c_m)$ is a solution of **(1.16)**.

Example 1.8

The differential equation

$$y'' - y' = 0$$

has the function f given by

$$f(x, c_1, c_2) = c_1 e^x + c_2$$

as a two-parameter family of solutions. To see this, take $\varphi(x) = f(x, c_1, c_2) = c_1 e^x + c_2$, and see that

$$\begin{aligned}
\varphi'(x) &= c_1 e^x; \\
\varphi''(x) &= c_1 e^x,
\end{aligned}$$

so

$$\varphi''(x) - \varphi'(x) = 0.$$

Hence, φ is a solution of $y'' - y' = 0$ no matter what values we choose for the constants c_1 and c_2.

As usual, we try to simplify our notation, and to refer to an n-parameter family of solutions f to a differential equation we will usually write $f(x)$ instead of $f(x, c_1, \ldots, c_n)$. Often, too, we will use the same symbol as that of the dependent variable of the differential equation to represent the family of solutions. For example, the family of solutions in example 1.8 may be represented by

$$y = c_1 e^x + c_2.$$

Most of the differential equations that arise in practice, as well as those that we will consider, have only one family of solutions, and such family of solutions has a number of parameters equal to the order of the differential equation. The reader should keep in mind, however, that this is not the case in general. For example, the differential equation of second order

$$|y''| + |y| = 0$$

has only one solution, namely the zero function $y \equiv 0$, and therefore it does not have a two-parameter family of solutions. Also, the number of parameters of a family of solutions to a differential equation of order n may be greater than n. For instance, the first order differential equation

$$(y')^2 - x^2 = 0$$

has

$$\left(y + \frac{1}{2}x^2 + c_1\right)\left(y - \frac{1}{2}x^2 + c_2\right) = 0$$

as a two-parameter family of (implicit) solutions. Note that we can equivalently say that this differential equation has the following two one-parameter families of solutions:

$$y_1 = -\frac{1}{2}x^2 + c_1 \quad \text{and} \quad y_2 = \frac{1}{2}x^2 + c_2.$$

As pointed out earlier, we will not consider differential equations of this kind, so we can assert here that a differential equation will always have one family of solutions whose number of parameters is equal to the order of the differential equation.

Definition 1.9

We say that the graph of a function φ is an **integral curve** of a differential equation if φ is a solution of that differential equation.

Example 1.10

Consider the differential equation

$$y' = x^2 + y. \tag{1.17}$$

The following are three solutions of this differential equation:

$$y_1(x) = e^x - x^2 - 2x - 2, \quad y_2(x) = -x^2 - 2x - 2, \quad y_3(x) = -2e^x - x^2 - 2x - 2.$$

Therefore, the graph of this functions are integral curves of (1.17) (see fig. 1.1).

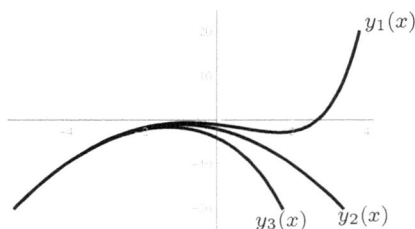

Figure 1.1. Integral curves corresponding to the solutions $y_1(x) = e^x - x^2 - 2x - 2$, $y_2(x) = -x^2 - 2x - 2$ and $y_3(x) = -2e^x - x^2 - 2x - 2$

Note that the one-parameter family of solutions

$$y = c_1 e^x - x^2 - 2x - 2$$

gives the solutions y_1, y_2, y_3 when choosing some values for c_1. Therefore, this family also determines a family of integral curves of the differential equation **(1.17)**.

1.1.4   Direction Fields

Let us now study the geometrical meaning of the assertion that y is a solution of the first order differential equation $y' = f(x, y)$. This discussion will lead to a geometrical technique for studying the behavior of the solutions of a first order differential equation. This is desirable because quite frequently it is impossible to analytically find the solution of a differential equation of the first order (let alone differential equations of order $n > 1$), and because in many applications a good qualitative understanding of a solution of a differential equation is sufficient.

First let us recall an idea that is well discussed in elementary calculus courses. Consider a function f differentiable in some interval I. If we draw any tangent to this graph at some point (x_1, y_1) within this interval, we will note that such tangent seems to overlap with the points of the graph of f in a small neighborhood centered at (x_1, y_1), because the points on the tangent line will be very close to the points on the graph in a small neighborhood of (x_1, y_1). Therefore, if we draw a small tangent segment to the graph of f at some point, this segment will actually appear as part of the graph of f. In figure 1.2 (a) we have drawn several tangent segments to a curve at different points. Note that these segments seem to overlap with the curve at almost every point of each segment. This means that we can use the little tangent segments by themselves to infer what the shape of the curve is. This fact is illustrated by part (b) of figure 1.2.

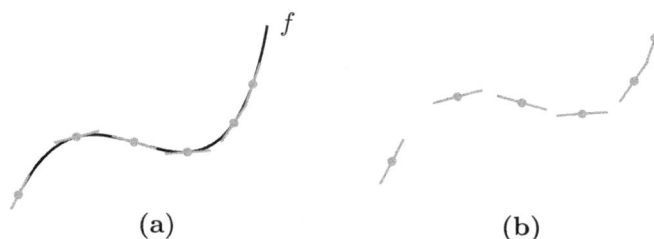

(a)   (b)

Figure 1.2. (a) Notice how the little segments in gray seem to overlap with the graph of f near the tangency point. (b) By looking at the little tangent segments we can actually imagine what the shape of the graph of f is like

The principle of similarity between the graph of a function f and its tangent line near the tangency point that we just discussed can be used to study the solutions of a first order differential equation from a geometrical perspective. When we say that $y(x)$ is a solution of the first order differential equation

$$y' = f(x, y),$$

we are actually claiming that if (a, b) is a point on the graph of $y(x)$, then the tangent line to the graph of $y(x)$ at this point has slope $f(a, b)$, because this gives us the value of the derivative of $y(x)$ at (a, b). Therefore, we can use $f(x, y)$ to compute the slope of little tangent segments at several points in a region R of the plane for which the differential equation has some solutions. From the discussion above, it follows that drawing enough of these little tangent segments, which are called **line elements**, will allow us to picture the shape of the integral curves in the region R. The diagram that results from drawing a set of these line elements is called a **direction field**.

Drawing a vector field by hand is certainly a tedious job, and we will not ask the student to perform such a task. Instead, we will tell you how to use *Mathematica* and the function `VectorPlot` to produce direction fields. Let's look at the following example.

Example 1.11

Use *Mathematica* to plot a direction field for the differential equation

$$y' = x + y,$$

in the region $R = [-5, 5] \times [-5, 5]$. Use the fact that $f(x) = ce^x - x - 1$ is a 1-parameter family of solutions to this DE to plot some integral curves of the DE in the direction field.

Solution: The function `VectorPlot` produces the plot of a vector field. To plot the direction field of a first order differential equation $y' = f(x, y)$ in the region $[a, b] \times [c, d]$, simply write

$$\texttt{VectorPlot[\{1,f(x,y)\},\{x,a,b\},\{y,c,d\}]}$$

We will now plot our direction field and assign the result to **dField**.

In[1]:= `dField = VectorPlot[Normalize[{1, x + y}],`
`{x, -5, 5}, {y, -5, 5}, VectorScale → 0.03]`

Out[1]:=

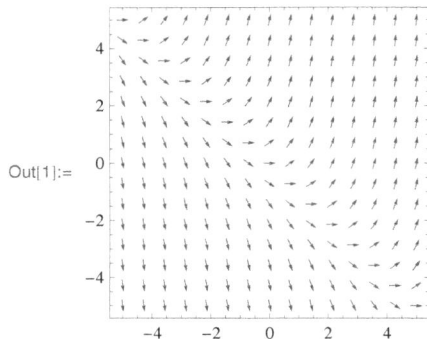

Note that **VectorPlot** will draw arrows instead of just line segments, since it is used to produce vector fields. The function **Normalize** is used to make all vectors have magnitude equal to 1, and the option **VectorScale** was used to reduce the size of the arrows.

Let's now sketch some integral curves in the direction field that we have obtained. First we will define the function for the family of solutions:

In[2]:= `y[x_, c_] := -1 - x + c e^x;`

Out[2]:=

Then we plot a few integral curves for $c = -10, -1, 0, 0.1, 0.5, 1$, and assign the result to **intCurves**.

In[3]:= `intCurves = Plot[(y[x, #1] &) /@ {-10, -1, 0, 0.1, 0.5, 1},`
`{x, -5, 5}, PlotStyle → Thick]`

Out[3]:=

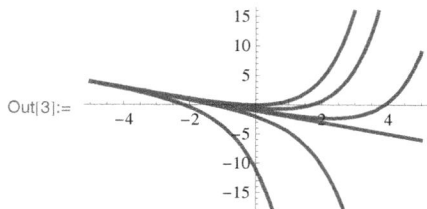

We can combine both graphs with **Show**:

In[4]:= `Show[{dField, intCurves}, Frame → None, Axes → True]`

Out[4]:=

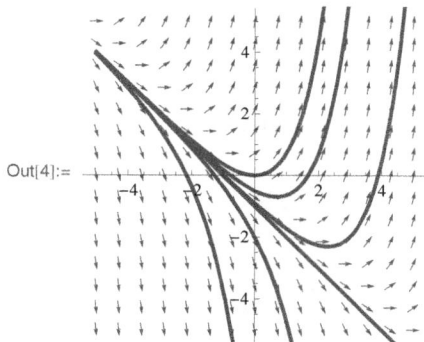

1.2 Some Linear Algebra

1.2.1 Matrices

Let us begin by introducing a mathematical concept that will provide us with an ideal notation for studying a variety of problems and ideas in linear algebra. Such concept is that of a **matrix**. We will begin by showing how a matrix is a convenient way of representing and handling a system of linear equations. Later, we will look at some matrix operations and derive a matrix algebra that will further enhance the usefulness of our matrix notation.

Matrices are rectangular arrays of (real or complex) numbers such as

$$\begin{pmatrix} -1 & 3 \\ 2 & 0 \\ 5 & 7 \end{pmatrix}, \quad \begin{pmatrix} 1 & -1 & -2i \\ 15 & x & 7 \end{pmatrix}, \quad \begin{pmatrix} 1 & -2 & 3 & 11 \\ 4 & 1 & -1 & 4 \\ 2 & -1 & 3 & 10 \end{pmatrix}.$$

In general, an $m \times n$ **matrix** $\mathbf{A}$ is a rectangular array of m rows and n columns such as

$$\mathbf{A} = \begin{pmatrix} a_{11} & a_{12} & \cdots & a_{1n} \\ a_{21} & a_{22} & \cdots & a_{2n} \\ \vdots & \vdots & \ddots & \vdots \\ a_{m1} & a_{m2} & \cdots & a_{mn} \end{pmatrix},$$

where each of the a_{ij}, with $1 \leq i \leq m$ and $1 \leq j \leq n$, is called an **entry** of the matrix $\mathbf{A}$. The entry in the i-th row and j-th column of a matrix $\mathbf{A}$ will be represented by a_{ij}, that is, by the corresponding lowercase letter together with a subindex ij. An $m \times n$ matrix is said to be of **size** $m \times n$.

Here we will use boldface letters, such as $\mathbf{A}$ or $\mathbf{B}$, to represent matrices.

Example 1.12

If **B** is the matrix

$$\mathbf{B} = \begin{pmatrix} 1 & -1 & -2i \\ 15 & x & 7 \end{pmatrix},$$

identify the entries b_{11}, b_{12}, b_{13}, b_{21}, b_{22} and b_{23}.

Solution: The entry b_{11} is the one in the first row and first column, so $b_{11} = 1$, while the entry b_{12} is the one in the first row and the second column, so $b_{12} = -1$. The rest of the indicated entries are $b_{13} = -2i$, $b_{21} = 15$, $b_{22} = x$ and $b_{23} = 7$.

A $n \times n$, which has as many rows as columns, is called a **square matrix of size n**.

A matrix **A** of size $1 \times n$ will be called a **row vector of n components**. Similarly, we will call a matrix **A** of size $n \times 1$ a **column vector of n components**. The following are a row vector of 5 components and a column vector of 4 components:

$$(-5, 3, 1, 0, -2) \qquad \text{and} \qquad \begin{pmatrix} 3 \\ -1 \\ 7 \\ 5 \end{pmatrix}.$$

If **A** is either a row vector or a column vector, each of its entries may also be referred to as a **component** of **A**. We will use lowercase boldface letters like **x** and **y** to represent vectors.

Note that an $m \times n$ matrix **A** can be seen as an arrangement

$$\begin{pmatrix} \mathbf{r}_1 \\ \mathbf{r}_2 \\ \vdots \\ \mathbf{r}_m \end{pmatrix}$$

of m row vectors of n components, as well as an arrangement

$$(\mathbf{c}_1 \ \mathbf{c}_2 \ \cdots \ \mathbf{c}_n)$$

of n column vectors of m components. For example, if

$$\mathbf{A} = \begin{pmatrix} 11 & 51 & 83 \\ -25 & -14 & 36 \\ 90 & 48 & 0 \\ -43 & 93 & 18 \end{pmatrix},$$

then we can say that

$$\mathbf{A} = \begin{pmatrix} \mathbf{r}_1 \\ \mathbf{r}_2 \\ \mathbf{r}_3 \\ \mathbf{r}_4 \end{pmatrix}$$

with $\mathbf{r}_1 = (11\ 51\ 83)$, $\mathbf{r}_2 = (-25\ -14\ 36)$, $\mathbf{r}_3 = (90\ 48\ 0)$ and $\mathbf{r}_4 = (-43\ 93\ 18)$. We can also see matrix $\mathbf{A}$ as the arrangement

$$\mathbf{A} = (\mathbf{c}_1\ \mathbf{c}_2\ \mathbf{c}_3)$$

with $\mathbf{c}_1 = \begin{pmatrix} 11 \\ -25 \\ 90 \\ -43 \end{pmatrix}$, $\mathbf{c}_2 = \begin{pmatrix} 51 \\ -14 \\ 48 \\ 93 \end{pmatrix}$, $\mathbf{c}_3 = \begin{pmatrix} 83 \\ 36 \\ 0 \\ 18 \end{pmatrix}$.

Representing a matrix in one of the above ways will be useful when discussing matrix operations.

If $\mathbf{A}$ is an $n \times n$, the **main diagonal** of $\mathbf{A}$ consists of all of its entries a_{ij} for which $i = j$. When the entries of $\mathbf{A}$ that are not in the main diagonal are all zero, we say that $\mathbf{A}$ is a **diagonal** matrix. For example, the following is a 5×5 diagonal matrix:

$$\begin{pmatrix} -1 & 0 & 0 & 0 & 0 \\ 0 & 4 & 0 & 0 & 0 \\ 0 & 0 & -2 & 0 & 0 \\ 0 & 0 & 0 & 7 & 0 \\ 0 & 0 & 0 & 0 & 3 \end{pmatrix}.$$

The particular $m \times n$ matrix whose entries are all zero is called the **zero matrix**, and it is denoted by $\mathbf{0}_{m \times n}$, or simply $\mathbf{0}$ whenever its size can be inferred from the context.

1.2.2   Systems of Linear Equations

Consider the system of linear equations

$$\begin{array}{rcl} x_1 - 2x_2 + 3x_3 &=& 11 \\ 4x_1 + x_2 - x_3 &=& 4 \\ 2x_1 - x_2 + 3x_3 &=& 10. \end{array} \tag{1.18}$$

This system consists of three equations in the three unknowns x_1, x_2 and x_3. To solve this system means finding all triplets (a, b, c) such that the substitution $x_1 = a$, $x_2 = b$ and $x_3 = c$ in (1.18) yields a true statement. Such a triplet is said to be a solution to system (1.18).

One way to solve system (1.18) is by eliminating unknown variables through addition or subtraction of the appropriate multiples of the equations of the system. For example, we can eliminate the unknown x_1 in the second equation by multiplying the first equation by -4 and adding the result to the second equation, which will yield

$$\begin{array}{rcl} x_1 - 2x_2 + 3x_3 &=& 11 \\ 9x_2 - 13x_3 &=& -40 \\ 2x_1 - x_2 + 3x_3 &=& 10. \end{array}$$

Multiplying the first equation by -2 and adding this to the third equation will eliminate x_1 from the third equation, resulting in the system

$$
\begin{aligned}
x_1 - 2x_2 + 3x_3 &= 11 \\
9x_2 - 13x_3 &= -40 \\
3x_2 - 3x_3 &= -12.
\end{aligned}
$$

Now we make the coefficient of x_2 in the second equation equal to 1 by multiplying this equation by $\frac{1}{9}$, obtaining

$$
\begin{aligned}
x_1 - 2x_2 + 3x_3 &= 11 \\
x_2 - \frac{13}{9}x_3 &= -\frac{40}{9} \\
3x_2 - 3x_3 &= -12,
\end{aligned}
$$

and multiplying the second equation by -3 and adding this to the third equation gives us

$$
\begin{aligned}
x_1 - 2x_2 + 3x_3 &= 11 \\
x_2 - \frac{13}{9}x_3 &= \frac{-40}{9} \\
\frac{4}{3}x_3 &= \frac{4}{3},
\end{aligned}
$$

From this we find $x_3 = 1$, and substituting this value for x_3 in the second equation shows that $x_2 = -3$. Finally, using the values $x_3 = 1$ and $x_2 = -3$ in the first equation we get $x_1 = 2$.

We can easily see that our process of solution of system **(1.18)** involved only operations with the coefficients of the unknown variables. Therefore, we can simplify our representation of the system by using a notation that records only the coefficients of the unknown variables in such a way that it is always clear which coefficient corresponds to what variable and to what equation of the system. We can do this by arranging the coefficients of system **(1.18)** in the matrix

$$
\begin{pmatrix}
1 & -2 & 3 \\
4 & 1 & -1 \\
2 & -1 & 3
\end{pmatrix}. \tag{1.19}
$$

Such a representation of **(1.18)** is called the **coefficient matrix** of the system **(1.18)**. If we append to this matrix a column containing the numbers in the right-hand side of the equations in **(1.18)**, we obtain

$$
\begin{pmatrix}
1 & -2 & 3 & 11 \\
4 & 1 & -1 & 4 \\
2 & -1 & 3 & 10
\end{pmatrix}, \tag{1.20}
$$

which is called the **augmented matrix** of system **(1.18)**.

Obtaining the solution to system **(1.18)** could then be achieved by performing operations between the rows of its augmented matrix, and eliminating variables from the equations of the system will correspond to introducing zeros in its augmented matrix. The row operations that we can do to solve the corresponding system are the following.

(1) Multiply the i-th row of an augmented matrix $\mathbf{A}$ by a constant c. This operation is denoted by cR_i. The notation $R_i \to cR_i$ will mean "replace the i-th row of the matrix by the multiple cR_i of itself".

(2) Substitute the i-th row by the sum of itself and a multiple of the j-th row. This operation will be denoted by $R_i \to R_i + cR_j$.

(3) Swap rows i and j. This operation will be represented by $R_i \rightleftarrows R_j$.

Performing any of the above row operations on an augmented matrix $\mathbf{A}$ yields an **equivalent matrix**, that is, a matrix that represents a system whose solution set is the same as that of the system represented by $\mathbf{A}$.

Let us know solve system **(1.18)** using our new notation. Eliminating x_1 from the second equation of **(1.18)** corresponds to making $a_{21} = 0$ in its augmented matrix $\mathbf{A}$. To do this, we replace R_2 by $-4R_1 + R_2$, an operation that can be represented by $R_2 \to -4R_1 + R_2$. Hence,

$$
\begin{pmatrix} 1 & -2 & 3 & 11 \\ 4 & 1 & -1 & 4 \\ 2 & -1 & 3 & 10 \end{pmatrix} \quad {\scriptstyle R_2 \to -4R_1+R_2} \quad \begin{pmatrix} 1 & -2 & 3 & 11 \\ 0 & 9 & -13 & -40 \\ 2 & -1 & 3 & 10 \end{pmatrix}.
$$

This is step is equivalent to the first step we carried out earlier to solve system **(1.18)**. We can continue and follow the procedure we used before to solve this system, this time using its augmented matrix. Doing this gives us the following sequence of row operations:

$$
\begin{pmatrix} 1 & -2 & 3 & 11 \\ 0 & 9 & -13 & -40 \\ 2 & -1 & 3 & 10 \end{pmatrix} \quad {\scriptstyle R_2 \to -2R_1+R_2} \quad \begin{pmatrix} 1 & -2 & 3 & 11 \\ 0 & 9 & -13 & -40 \\ 0 & 3 & -3 & -12 \end{pmatrix}
$$

$$
{\scriptstyle R_2 \to \frac{1}{9}R_2} \quad \begin{pmatrix} 1 & -2 & 3 & 11 \\ 0 & 1 & -\frac{13}{9} & -\frac{40}{9} \\ 0 & 3 & -3 & -12 \end{pmatrix}
$$

$$
{\scriptstyle R_3 \to -3R_2+R_3} \quad \begin{pmatrix} 1 & -2 & 3 & 11 \\ 0 & 1 & -\frac{13}{9} & -\frac{40}{9} \\ 0 & 0 & \frac{4}{3} & \frac{4}{3} \end{pmatrix}
$$

$$
{\scriptstyle R_3 \to \frac{3}{4}R_3} \quad \begin{pmatrix} 1 & -2 & 3 & 11 \\ 0 & 1 & -\frac{13}{9} & -\frac{40}{9} \\ 0 & 0 & 1 & 1 \end{pmatrix}.
$$

The last matrix is the augmented matrix of the system

$$
\begin{aligned}
x_1 - 2x_2 + 3x_3 &= 11 \\
x_2 - \tfrac{13}{9}x_3 &= \tfrac{-40}{9} \\
x_3 &= 1,
\end{aligned}
$$

Obtaining the solution is now straightforward. As before, the solution is $(1, -3, 2)$.

We see then that solving a system of n linear equations in n unknowns can be done by doing row operations in its augmented matrix. We will perform these operations until we get a matrix whose entries below the main diagonal are zeros (such a matrix is said to be in **row echelon form**).

Let us now stop our discussion on systems of linear equations and introduce more of the theory of matrices. The concepts we will study will allow us to write facts about systems of linear equations using concise expressions.

1.2.3   Matrix Addition and Subtraction and Scalar Product

Definition 1.13

Let $\mathbf{A}$ and $\mathbf{B}$ be two $m \times n$ matrices. The **sum** of $\mathbf{A}$ and $\mathbf{B}$ is a matrix $\mathbf{C}$ obtained from adding $\mathbf{A}$ and $\mathbf{B}$ entry-wise, that is, $\mathbf{C}$ is the matrix whose entry c_{ij} is the sum $a_{ij} + b_{ij}$. If α is any number, we also define the **scalar product** of α and $\mathbf{A}$, denoted $\alpha\mathbf{A}$, as the matrix whose ij entry is the product αa_{ij}. Hence, $\alpha\mathbf{A}$ is obtained by multiplying each entry of $\mathbf{A}$ by the number α.

Remark

Note that the sum of two matrices is only defined for matrices of the same size, so if two matrices differ by their number of rows or by their number of columns, they cannot be added. For example, the sum

$$\begin{pmatrix} -3 & 3 \\ 2 & 0 \\ 5 & 7 \end{pmatrix} + \begin{pmatrix} 1 & -1 & -3 \\ 5 & 2 & 4 \end{pmatrix}$$

cannot be performed.

Example 1.14

Let

$$\mathbf{A} = \begin{pmatrix} 3 & -2 & 1 \\ 15 & 1 & 0 \end{pmatrix} \quad \text{and} \quad \mathbf{B} = \begin{pmatrix} 7 & 2 & 5 \\ -1 & -3 & 9 \end{pmatrix}.$$

Compute the sum $\mathbf{C} = \mathbf{A} + \mathbf{B}$ and the products $-2\mathbf{A}$ and $3\mathbf{B}$. The sum $\mathbf{C}$ is obtained by adding the matrices $\mathbf{A}$ and $\mathbf{B}$ entry-wise, so the entry c_{11} is the sum $a_{11} + b_{11}$, the entry c_{12} is the sum $a_{12} + b_{12}$, and so forth. Hence, we have

$$
\begin{aligned}
c_{11} &= 3 + 7 = 10, \\
c_{12} &= -2 + 2 = 0, \\
c_{13} &= 1 + 5 = 6, \\
c_{21} &= 15 - 1 = 14, \\
c_{22} &= 1 - 3 = -2, \\
c_{23} &= 0 + 9 = 9,
\end{aligned}
$$

so

$$
\mathbf{A} + \mathbf{B} = \begin{pmatrix} 10 & 0 & 6 \\ 14 & -2 & 9 \end{pmatrix}.
$$

Computing $-2\mathbf{A}$ is done by multiplying each entry of $\mathbf{A}$ by -2, so

$$
-2\mathbf{A} = \begin{pmatrix} -6 & 4 & -2 \\ -30 & -2 & 0 \end{pmatrix}.
$$

Similarly,

$$
3\mathbf{B} = \begin{pmatrix} 21 & 6 & 15 \\ -3 & -9 & 27 \end{pmatrix}.
$$

The **difference** of two matrices $\mathbf{A}$ and $\mathbf{B}$ is defined as the matrix

$$
\mathbf{A} - \mathbf{B} = \mathbf{A} + (-\mathbf{B}).
$$

This means that $\mathbf{A} - \mathbf{B}$ is obtained from subtracting each entry in $\mathbf{B}$ from the corresponding entry in $\mathbf{A}$.

Example 1.15

For the matrices $\mathbf{A}$ and $\mathbf{B}$ from example 1.14, we have

$$
\mathbf{A} - \mathbf{B} = \begin{pmatrix} 3 & -2 & 1 \\ 15 & 1 & 0 \end{pmatrix} - \begin{pmatrix} 7 & 2 & 5 \\ -1 & -3 & 9 \end{pmatrix} = \begin{pmatrix} 3-7 & -2-2 & 1-5 \\ 15-(-1) & 1-(-3) & 0-9 \end{pmatrix}
$$
$$
= \begin{pmatrix} -4 & -4 & -4 \\ 16 & 4 & -9 \end{pmatrix}.
$$

Before going into the definition of the product of two matrices, it is convenient to introduce some concepts that will allow us to formulate the definition in a nice and clean way. This concepts will also be useful in other situations that we will encounter later.

1.2.4   Transpose of a Matrix

The first concept we will look at is that of the **transpose** of a matrix.

Definition 1.16

Let $\mathbf{A}$ be an $m \times n$ matrix. The **transpose** of $\mathbf{A}$, denoted $\mathbf{A}^T$, is the matrix that results from turning the rows of $\mathbf{A}$ into columns. More precisely, if

$$\mathbf{A} = \begin{pmatrix} a_{11} & a_{12} & \cdots & a_{1n} \\ a_{21} & a_{22} & \cdots & a_{2n} \\ \vdots & \vdots & \ddots & \vdots \\ a_{m1} & a_{m2} & \cdots & a_{mn} \end{pmatrix},$$

then

$$\mathbf{A}^T = \begin{pmatrix} a_{11} & a_{21} & \cdots & a_{m1} \\ a_{12} & a_{22} & \cdots & a_{m2} \\ \vdots & \vdots & \ddots & \vdots \\ a_{1n} & a_{2n} & \cdots & a_{mn} \end{pmatrix}.$$

Example 1.17

Let $\mathbf{A} = \begin{pmatrix} 1 & 4 \\ -2 & 5 \end{pmatrix}$ and $\mathbf{B} = \begin{pmatrix} -1 & 3 & 5 & 1 \\ 4 & -5 & 7 & 10 \end{pmatrix}$. Compute $\mathbf{A}^T$ and $\mathbf{B}^T$.

Solution: Interchanging rows for columns in matrix $\mathbf{A}$ gives us

$$\mathbf{A}^T = \begin{pmatrix} 1 & -2 \\ 4 & 5 \end{pmatrix}.$$

For matrix $\mathbf{B}$ we have

$$\mathbf{B}^T = \begin{pmatrix} -1 & 4 \\ 3 & -5 \\ 5 & 7 \\ 1 & 10 \end{pmatrix}.$$

Note that the transpose of any row vector is a column vector with the same number of components.

1.2.5   Dot Product and Matrix Multiplication

We will now define a special type of product between two vectors. This operation will be used to formulate a more general product defined for matrices of other sizes.

Definition 1.18

Let

$$\mathbf{x} = \begin{pmatrix} x_1 \\ x_2 \\ \vdots \\ x_n \end{pmatrix} \qquad \text{and} \qquad \mathbf{y} = \begin{pmatrix} y_1 \\ y_2 \\ \vdots \\ y_n \end{pmatrix}$$

The **dot product** of **x** and **y**, denoted by $x \cdot y$, is the number

$$\mathbf{x} \cdot \mathbf{y} = x_1 y_1 + x_2 y_2 + \cdots + x_n y_n,$$

or, using the sigma notation for a summation,

$$\mathbf{x} \cdot \mathbf{y} = \sum_{i=1}^{n} x_i y_i.$$

The dot product $\mathbf{x} \cdot \mathbf{y}$ for row vectors **x** and **y** can then be defined as $\mathbf{x} \cdot \mathbf{y} = \mathbf{x}^T \cdot \mathbf{y}^T$, that is, we turn **x** and **y** into column vectors and compute the dot product. It is convenient to define the dot product of vectors that are not both row vectors or column vectors, but still have the same number of components. In this case we can define $\mathbf{x} \cdot \mathbf{y} = \mathbf{x}^T \cdot \mathbf{y}$, where the vectors in the right-hand side are both either row vectors or column vectors.

> **Remark**
> Note that the dot product is defined only for vectors of the same size. Moreover, we see that the dot product of two vectors is just a number, not a new vector.

Example 1.19

Let

$$\mathbf{a} = \begin{pmatrix} 10 \\ 8 \\ -5 \\ 0 \end{pmatrix}, \qquad \mathbf{b} = \begin{pmatrix} 5 \\ 9 \\ -4 \\ -4 \end{pmatrix}, \qquad \mathbf{c} = (-7 \ -12 \ -15 \ 5 \ -8), \qquad \mathbf{d} = \begin{pmatrix} 15 \\ 12 \\ -6 \\ -7 \\ 15 \end{pmatrix}.$$

Compute $\mathbf{a} \cdot \mathbf{b}$ and $\mathbf{c} \cdot \mathbf{d}$.

Solution: From definition 1.18, we have

$$\mathbf{a} \cdot \mathbf{b} = (10)(5) + (8)(9) + (-5)(-4) + (0)(-4) = 142.$$

The dot product $\mathbf{c} \cdot \mathbf{d}$ is computed similarly:

$$\mathbf{c} \cdot \mathbf{d} = (-7)(12) + (-12)(12) + (-15)(-6) + (5)(-7) + (-8)(15) = -314.$$

We will know define the matrix multiplication.

Definition 1.20

Let **A** be an $m \times n$ matrix and **B** be an $n \times p$ matrix, and consider the representations

$$\mathbf{A} = \begin{pmatrix} \mathbf{a}_1 \\ \mathbf{a}_2 \\ \vdots \\ \mathbf{a}_m \end{pmatrix} \qquad \mathbf{B} = (\mathbf{b}_1 \ \mathbf{b}_2 \ \cdots \ \mathbf{b}_p),$$

where each of the $\mathbf{a}_i$ is a row of $\mathbf{A}$ (of n components) and each of the $\mathbf{b}_j$ is a column of $\mathbf{B}$ (of n components). The **product** of $\mathbf{A}$ and $\mathbf{B}$, denoted $\mathbf{AB}$, is the matrix $m \times p$ $\mathbf{C}$ whose ij entry is given by

$$c_{ij} = \mathbf{a}_i \cdot \mathbf{b}_j = \sum_{k=1}^{n} a_{ik}b_{kj},$$

that is, c_{ij} is the (dot) product of row i of $\mathbf{A}$ and column j of $\mathbf{B}$.

Remark
From the definition above, we see that the product $\mathbf{AB}$ of two matrices $\mathbf{A}$ and $\mathbf{B}$ is defined only when the number of columns of $\mathbf{A}$ equals the number of rows of $\mathbf{B}$. Otherwise, a row of $\mathbf{A}$ and a column of $\mathbf{B}$ will not have the same number of components, and thus their dot product is not defined.

Example 1.21

Let

$$\mathbf{A} = \begin{pmatrix} 3 & 10 & 1 & 0 \\ 2 & 6 & -8 & -2 \\ 7 & 2 & 7 & -8 \end{pmatrix} \quad \text{and} \quad \mathbf{B} = \begin{pmatrix} 5 & -10 & -10 & -1 \\ 4 & -9 & 6 & 4 \\ 2 & -5 & 9 & -4 \\ -5 & 9 & 9 & -9 \end{pmatrix}.$$

If defined, compute the product $\mathbf{C} = \mathbf{AB}$ and $\mathbf{D} = \mathbf{BA}$.

Solution: Let us start by checking that matrices $\mathbf{A}$ and $\mathbf{B}$ have the proper size for the products $\mathbf{C} = \mathbf{AB}$ and $\mathbf{D} = \mathbf{BA}$ to be defined. We see that $\mathbf{A}$ has as many columns as $\mathbf{B}$ has rows, so $\mathbf{C} = \mathbf{AB}$ is defined. However, $\mathbf{B}$ has 4 columns and $\mathbf{A}$ has 3 rows, so the second product $\mathbf{D} = \mathbf{BA}$ is not defined. Therefore, we will only compute $\mathbf{C} = \mathbf{AB}$.

We begin by computing the entry c_{11}. This is the dot product of the first row of $\mathbf{A}$ and the first column of $\mathbf{B}$, so

$$c_{11} = (3 \ \ 10 \ \ 1 \ \ 0) \cdot \begin{pmatrix} 5 \\ 4 \\ 2 \\ -5 \end{pmatrix} = 57.$$

We can continue and compute the entries c_{12}, c_{13} and c_{14} to have the first row of matrix $\mathbf{C}$, but we leave this computation to the reader. Instead, we will compute the entry c_{24}, which is the dot product of row 2 of matrix $\mathbf{A}$ and column 4 of matrix $\mathbf{B}$. We have

$$c_{24} = (2 \ \ 6 \ \ -8 \ \ -2) \cdot \begin{pmatrix} -1 \\ 4 \\ -4 \\ -9 \end{pmatrix} = 72.$$

We leave the computation of the remaining entries of $\mathbf{C}$ to the reader, which gives the 3×4 product matrix

$$\mathbf{C} = \begin{pmatrix} 57 & -125 & 39 & 33 \\ 28 & -52 & -74 & 72 \\ 97 & -195 & -67 & 45 \end{pmatrix}.$$

If $\mathbf{A}$ is an $n \times n$ matrix, we can multiply $\mathbf{A}$ by itself, and we can therefore define positive integer powers of $\mathbf{A}$ just by

$$\mathbf{A}^n = \underbrace{\mathbf{A}\mathbf{A}\cdots\mathbf{A}}_{n \text{ factors}}.$$

If D is an $n \times n$ diagonal matrix, computing D^k only requires rising each entry on the diagonal of D to the k-th power. That is,

$$D = \begin{pmatrix} d_1 & 0 & 0 & \cdots & 0 \\ 0 & d_2 & 0 & \cdots & 0 \\ 0 & 0 & d_3 & & 0 \\ \vdots & \vdots & & \ddots & \vdots \\ 0 & 0 & 0 & \cdots & d_n \end{pmatrix} \quad \Rightarrow \quad D^k = \begin{pmatrix} d_1^k & 0 & 0 & \cdots & 0 \\ 0 & d_2^k & 0 & \cdots & 0 \\ 0 & 0 & d_3^k & & 0 \\ \vdots & \vdots & & \ddots & \vdots \\ 0 & 0 & 0 & \cdots & d_n^k \end{pmatrix}.$$

Here are some algebraic properties of matrix addition and multiplication.

Theorem 1.22

Let $\mathbf{A}$, $\mathbf{B}$ and $\mathbf{C}$ be matrices whose dimensions we will suppose appropriate to perform matrix addition and matrix multiplication in the statements below. The following properties hold.

(1) $\mathbf{A} + \mathbf{B} = \mathbf{B} + \mathbf{A}$;

(2) $\mathbf{A} + (\mathbf{B} + \mathbf{C}) = (\mathbf{A} + \mathbf{B}) + \mathbf{C}$.

(3) $\mathbf{A}(\mathbf{B}\mathbf{C}) = (\mathbf{A}\mathbf{B})\mathbf{C}$.

(4) $\mathbf{A}(\mathbf{B} + \mathbf{C}) = \mathbf{A}\mathbf{B} + \mathbf{A}\mathbf{C}$ *and* $(\mathbf{A} + \mathbf{B})\mathbf{C} = \mathbf{A}\mathbf{C} + \mathbf{B}\mathbf{C}$;

1.2.6   Determinants

There is a number associated to every *square* matrix $\mathbf{A}$, called the **determinant** of $\mathbf{A}$ and denoted $\det(\mathbf{A})$ or $|\mathbf{A}|$. For a 2×2 matrix

$$\mathbf{A} = \begin{pmatrix} a_{11} & a_{12} \\ a_{21} & a_{22} \end{pmatrix}$$

the determinant is defined as

$$\begin{vmatrix} a_{11} & a_{12} \\ a_{21} & a_{22} \end{vmatrix} = a_{11}a_{22} - a_{12}a_{21}. \tag{1.21}$$

If $\mathbf{A}$ is a 3×3 matrix, say

$$\begin{pmatrix} a_{11} & a_{12} & a_{13} \\ a_{21} & a_{22} & a_{23} \\ a_{31} & a_{32} & a_{33} \end{pmatrix},$$

then the determinant of $\mathbf{A}$ is computed as follows

$$\begin{vmatrix} a_{11} & a_{12} & a_{13} \\ a_{21} & a_{22} & a_{23} \\ a_{31} & a_{32} & a_{33} \end{vmatrix} = a_{11} \begin{vmatrix} a_{22} & a_{23} \\ a_{32} & a_{33} \end{vmatrix} - a_{12} \begin{vmatrix} a_{21} & a_{23} \\ a_{31} & a_{33} \end{vmatrix} + a_{13} \begin{vmatrix} a_{21} & a_{22} \\ a_{31} & a_{32} \end{vmatrix}. \quad (1.22)$$

In general, the determinant of a square matrix can be computed using a **cofactor expansion**. However, we will be dealing only with determinants of 2×2 or 3×3 matrices, so the formulas given above for the computation of determinants is all that we need.

Example 1.23

Let

$$\mathbf{A} = \begin{pmatrix} -4 & 6 \\ -5 & 9 \end{pmatrix} \quad \text{and} \quad \mathbf{B} = \begin{pmatrix} 1 & 1 & -5 \\ 8 & -3 & 3 \\ 4 & -6 & -5 \end{pmatrix}.$$

Compute $|\mathbf{A}|$ and $|\mathbf{B}|$.

Solution: From **(1.21)**, we have

$$|\mathbf{A}| = (-4)(9) - (6)(-5) = -6.$$

To compute $|\mathbf{B}|$, we could use formula **(1.22)** (which is a cofactor expansion of a 3×3 determinant). However, we will use this example to introduce a procedure that might be easier to remember. First, we write $|\mathbf{B}|$ and append a copy of the first two rows to the left:

$$\begin{vmatrix} 1 & 1 & -5 \\ 8 & -3 & 3 \\ 4 & -6 & -5 \end{vmatrix} \begin{matrix} 1 & 1 \\ 8 & -3. \\ 4 & -6 \end{matrix}$$

Each of the arrows in the picture above goes through a diagonal of three entries in the matrix. To compute the determinant, we only have to multiply the entries in each of these diagonals, and add or subtract the products that we obtain: if the diagonal goes from left to right, we add, and if the diagonal goes from right to left, we subtract. Hence,

$$\begin{aligned} |\mathbf{B}| &= (1)(-3)(-5) + (1)(3)(4) + (-5)(8)(-6) - (-5)(-3)(4) - (1)(3)(-6) \\ &\quad - (1)(8)(-5) \\ &= 265. \end{aligned}$$

Now some facts about determinants.

Theorem 1.24

Let **A** *and* **B** *be two* $n \times n$ *matrices.*

(1) $\det(\mathbf{AB}) = \det(\mathbf{A})\det(\mathbf{B})$.

(2) $\det(\mathbf{A}^T) = \det(\mathbf{A})$.

(3) *If* **B** *is the result of applying the operation* $R_i \rightleftarrows R_j$ *(swapping two rows) to* **A**, *then* $\det(\mathbf{B}) = -\det(\mathbf{A})$.

(4) *If* **B** *is the result of applying the operation* $R_i \to cR_i$ *to* **A**, *then* $\det(\mathbf{B}) = c\det(\mathbf{A})$.

(5) *If* **B** *is the result of applying the operation* $R_i \to R_i + cR_j$ *to* **A**, *then* $\det(\mathbf{B}) = \det(\mathbf{A})$.

(6) *Items (3)–(5) also hold if the operations are done with columns of* **A** *instead of rows.*

(7) $\det(\mathbf{A}) = 0$ *if any of the following is true:*

  (a) *there is one row (resp. column) whose components are all zeros;*

  (b) **A** *has two rows (resp. columns) that are equal to each other;*

  (c) **A** *has a row (resp. column) that is the multiple of another row (resp. column)*

1.2.7   The Inverse of a Square Matrix

If **A** is an $n \times n$ matrix, the sequence of all entries a_{ij} of **A** such that $i = j$ is called the **main diagonal** of **A**. The $n \times n$ matrix that contains only 1s in the main diagonal and 0s anywhere else is called the **identity matrix of size** $\boldsymbol{n}$, and it is denoted $\mathbf{I}_n$, or simply **I** if its size does not need to be mentioned explicitly. For example,

$$\mathbf{I}_4 = \begin{pmatrix} 1 & 0 & 0 & 0 \\ 0 & 1 & 0 & 0 \\ 0 & 0 & 1 & 0 \\ 0 & 0 & 0 & 1 \end{pmatrix}.$$

If **A** is an $n \times n$ matrix, then

$$\mathbf{AI}_n = \mathbf{I}_n\mathbf{A} = \mathbf{A}.$$

The equality above is the reason the matrix $\mathbf{I}_n$ is called the identity matrix.

Let **A** and **B** be two $n \times n$ matrices, and suppose that

$$\mathbf{AB} = \mathbf{BA} = \mathbf{I}_n.$$

Then **B** is called the **inverse** of **A**, and it is denoted $\mathbf{A}^{-1}$. If a matrix **A** has an inverse, we say that **A** is invertible. A matrix that is not invertible is also called a **singular** matrix, wheras an invertible matrix is also called a **nonsingular** matrix.

Here are some facts regarding invertible matrices.

Theorem 1.25

(1) *If* $\mathbf{A}$ *is invertible, then its inverse is unique. Clearly* $(\mathbf{A}^{-1})^{-1} = \mathbf{A}$.

(2) *If* $\mathbf{A}$ *and* $\mathbf{B}$ *are invertible, then* $\mathbf{AB}$ *is also invertible. Moreover,* $(\mathbf{AB})^{-1} = \mathbf{B}^{-1}\mathbf{A}^{-1}$. *Do not confuse this equality with* $(\mathbf{AB})^{-1} = \mathbf{A}^{-1}\mathbf{B}^{-1}$, *which is not true in general.*

(3) *If* $\mathbf{A}$ *is invertible, then* $\det(\mathbf{A}^{-1}) = \frac{1}{\det(\mathbf{A})}$.

(4) *A matrix* $\mathbf{A}$ *is invertible if and only if* $\det(\mathbf{A}) \neq 0$. *Equivalently,* $\mathbf{A}$ *is a singular matrix if and only if* $\det(\mathbf{A}) = 0$.

1.2.8   Matrix Form of a System of Linear Equations

Consider a system of n linear equations in n unknowns,

$$
\begin{aligned}
a_{11}x_1 + a_{12}x_2 + \cdots + a_{1n}x_n &= b_1 \\
\vdots \qquad\qquad \vdots \qquad \vdots & \\
a_{n1}x_1 + a_{n2}x_2 + \cdots + a_{nn}x_n &= b_n
\end{aligned}
\tag{1.23}
$$

Note that we can represent the system, using column vectors, as

$$
\begin{pmatrix}
a_{11}x_1 & + & a_{12}x_2 & + & \cdots & + & a_{1n}x_n \\
a_{21}x_1 & + & a_{22}x_2 & + & \cdots & + & a_{2n}x_n \\
\vdots & & & & \vdots & & \\
a_{n1}x_1 & + & a_{n2}x_2 & + & \cdots & + & a_{nn}x_n
\end{pmatrix}
=
\begin{pmatrix} b_1 \\ b_2 \\ \vdots \\ b_n \end{pmatrix}.
$$

The left-hand side of this equation can be written as the product

$$
\begin{pmatrix}
a_{11} & a_{12} & \cdots & a_{1n} \\
a_{21} & a_{22} & \cdots & a_{2n} \\
\vdots & & & \vdots \\
a_{n1} & a_{n2} & \cdots & a_{nn}
\end{pmatrix}
\begin{pmatrix} x_1 \\ x_2 \\ \vdots \\ x_n \end{pmatrix}.
$$

Hence, the system can be written as

$$
\mathbf{Ax} = \mathbf{b}, \tag{1.24}
$$

where $\mathbf{A}$ is the coefficient matrix of (**1.23**) and $\mathbf{x}$ and $\mathbf{b}$ are the vectors whose components are the unknowns and the values in the right hand side of this system, respectively.

Using the matrix representation of a system, we can concisely formulate some facts about systems of linear equations.

Theorem 1.26

Let $\mathbf{Ax} = \mathbf{b}$ *be a system of* n *linear equations in* n *unknowns. If* $\mathbf{A}$ *is nonsingular, then the system has a unique solution, whereas if* $\mathbf{A}$ *is singular, then the system has either no solution or has infinitely many solutions. In particular, if the system is homogeneous, i.e. if* $\mathbf{b} = \mathbf{0}$, *and* $\mathbf{A}$ *is singular, then the system has infinitely many solutions.*

1.2.9   Linear Dependence

Definition 1.27

Let $\mathbf{x}_1, \mathbf{x}_2, \ldots, \mathbf{x}_n$ be vectors, all with the same number of components, and let $S = \{\mathbf{x}_1, \mathbf{x}_2, \ldots, \mathbf{x}_n\}$. We say that the set S is **linearly dependent** if there exist constants $c_1, c_2, \ldots c_n$, not all of them zero, such that

$$c_1\mathbf{x}_1 + c_2\mathbf{x}_2 + \cdots + c_n\mathbf{x}_n = \mathbf{0}. \tag{1.25}$$

If the equation **(1.25)** holds only when all of the constants $c_1, \ldots, c_n$ are zero, then we say that the set S is **linearly independent**. That is, S is linearly independent if and only if

$$c_1\mathbf{x}_1 + c_2\mathbf{x}_2 + \cdots + c_n\mathbf{x}_n = \mathbf{0} \qquad \text{implies} \qquad c_i = 0$$

for all $i \in \{1, 2, \ldots, n\}$.

If all of the n vectors in a set $S = \{\mathbf{x}_1, \mathbf{x}_2, \ldots, \mathbf{x}_n\}$ have exactly n components, we can easily establish conditions upon which the set S is linearly dependent or linearly independent. If each $\mathbf{x}_j$ in S has n components, then equation **(1.25)** can be written as

$$c_1 \begin{pmatrix} x_{11} \\ x_{21} \\ \vdots \\ x_{n1} \end{pmatrix} + c_2 \begin{pmatrix} x_{12} \\ x_{22} \\ \vdots \\ x_{n2} \end{pmatrix} + \cdots + c_n \begin{pmatrix} x_{1n} \\ x_{2n} \\ \vdots \\ x_{nn} \end{pmatrix} = \mathbf{0}. \tag{1.26}$$

Finding all constants $c_1, c_2, \ldots$ for which equation **(1.26)** holds is equivalent to solving the homogeneous system of equations

$$\begin{aligned} x_{11}c_1 + x_{12}c_2 + \cdots + x_{1n}c_n &= 0 \\ \vdots \qquad\qquad \vdots \qquad\quad \vdots \\ x_{n1}c_1 + x_{n2}c_2 + \cdots + x_{nn}c_n &= 0. \end{aligned}$$

We know that this system can be represented as the matrix equation

$$\mathbf{Ac} = \mathbf{0}, \tag{1.27}$$

where $\mathbf{A}$ is the matrix whose columns are the vectors in S, that is, $\mathbf{A} = (\mathbf{x}_1 \ \mathbf{x}_2 \ \ldots \ \mathbf{x}_n)$, and $\mathbf{c}$ is the column vector whose entries are the unknown constants $c_1, c_2, \ldots, c_n$. From theorems 1.25 and 1.26, we know that the system represented by (1.27) has nontrivial solutions if and only if $\det(\mathbf{A}) = 0$. Consequently, the set S is linearly dependent if and only if $\det(\mathbf{A}) = 0$. Equivalently, we can say that the set S is linearly independent if and only if $\det(\mathbf{A}) \neq 0$. Let us state this result as a theorem.

Theorem 1.28

Let $S = \{\mathbf{x}_1, \mathbf{x}_2, \ldots, \mathbf{x}_n\}$ be a set of n vectors with n components, and let $\mathbf{A}$ be the matrix $\mathbf{A} = (\mathbf{x}_1\ \mathbf{x}_2\ \cdots\ \mathbf{x}_n)$. Then the set S is linearly dependent if and only if

$$\det(\mathbf{A}) = 0.$$

Equivalently, the set S is linearly independent if and only if $\det(\mathbf{A}) \neq 0$.

Example 1.29

Let

$$\mathbf{x}_1 = \begin{pmatrix} -1 \\ 0 \\ 2 \end{pmatrix}, \qquad \mathbf{x}_2 = \begin{pmatrix} 0 \\ -2 \\ 1 \end{pmatrix}, \qquad \mathbf{x}_3 = \begin{pmatrix} 2 \\ 1 \\ 3 \end{pmatrix}.$$

Determine whether the set $S = \{\mathbf{x}_1, \mathbf{x}_2, \mathbf{x}_3\}$ is linearly dependent or linearly independent.

Solution: We form the matrix

$$\mathbf{A} = (\mathbf{x}_1\ \ \mathbf{x}_2\ \ \mathbf{x}_3) = \begin{pmatrix} -1 & 0 & 1 \\ 0 & -2 & 3 \\ 2 & 1 & 3 \end{pmatrix},$$

and we compute the determinant of $\mathbf{A}$,

$$\det(\mathbf{A}) = 6 + 0 + 0 - (-4) - (-3) - 0 = 13 \neq 0.$$

Hence, the set $S = \{\mathbf{x}_1, \mathbf{x}_2, \mathbf{x}_3\}$ is linearly independent.

We can also define linear dependence and linear independence for vectors whose components are functions, that is, vectors of the form

$$\mathbf{x}(t) = \begin{pmatrix} x_1(t) \\ x_2(t) \\ \vdots \\ x_n(t) \end{pmatrix},$$

where each of the components $x_i(t)$ is a function of a variable t.

Definition 1.30

Let $S = \{\mathbf{x}_1(t), \mathbf{x}_2(t), \ldots, \mathbf{x}_n(t)\}$, where each vector $\mathbf{x}_j(t)$ of S is a vector function. Suppose

$$\mathbf{x}_j(t) = \begin{pmatrix} x_{1j}(t) \\ x_{2j}(t) \\ \vdots \\ x_{nj} \end{pmatrix},$$

and that each of the components $x_{ij}(t)$ is defined in an interval $[a,b]$ for all $i,j \in \{1,2,\ldots,n\}$. We say that the set S is **linearly dependent** in $[a,b]$ if and only if there exist constants $c_1, c_2, \ldots, c_n$, not all of them zero, such that

$$c_1\mathbf{x}_1(t) + c_2\mathbf{x}_2(t) + \cdots + c_n\mathbf{x}_n(t) = \mathbf{0} \qquad \text{for all } t \in [a,b].$$

Otherwise, the set S is said to be **linearly independent**. That is, S is linearly independent if and only if

$$c_1\mathbf{x}_1(t) + c_2\mathbf{x}_2(t) + \cdots + c_n\mathbf{x}_n(t) = \mathbf{0} \quad \text{for all } t \in [a,b]$$

implies $c_i = 0$ for all $i \in \{1,2,\ldots,n\}$.

Often, we will also discuss linear dependence of a set of functions, such as $S = \{\cos x, \sin x\}$. To deal with this kind of sets we do not need additional definitions, since functions themselves can be seen as vectors. The following is an example of a linearly independent set of two functions.

Example 1.31

Show that the set $S = \{e^x, e^{-x}\}$ is linearly independent on $(-\infty, \infty)$.

Solution: To determine if S is linearly dependent or linearly independent, definition 1.30 tells us that we need to determine if there are constants c_1 and c_2, not all of them zero, such that

$$c_1 e^x + c_2 e^{-x} = 0. \qquad (1.28)$$

If we multiply the above equation by e^x, we obtain

$$c_1 e^{2x} + c_2 = 0,$$

so

$$c_1 e^{2x} = -c_2.$$

If the left-hand side, $c_1 e^{2x}$, is a constant function, then it can only be $c_1 e^{2x} = 0$ for all $t \in (-\infty, \infty)$, which happens only by making $c_1 = 0$, since e^{2x} is never zero. Hence, the only constants c_1 and c_2 that satisfy (1.28) are $c_1 = 0$ and $c_2 = 0$, showing that $S = \{e^x, e^{-x}\}$ is linearly independent.

At this point it is a good idea to ask the reader to verify that the sets $\{x, x^2\}$ and $\{\sin x, \cos x\}$ are linearly independent on the whole real line.

We can also establish a result similar to theorem 1.28 to determine when a set of vectors is linearly independent. Note, however, that there are some important differences, which we will discuss after stating the theorem.

Theorem 1.32

Let $S = \{\mathbf{x}_1(t), \mathbf{x}_2(t), \ldots, \mathbf{x}(t)_n\}$ be a set of n vector functions each with n components, and let $\mathbf{A}(t)$ be the matrix function $\mathbf{A}(t) = (\mathbf{x}_1(t) \ \mathbf{x}_2(t) \ \ldots \ \mathbf{x}(t)_n)$, that is, the matrix whose j-th column is the vector $\mathbf{x}_j$ of S. If the set S is linearly dependent on the interval $[a,b]$, then $\det(\mathbf{A}(t)) = 0$ for all $t \in [a,b]$. Equivalently, if $\det(\mathbf{A}(t)) \neq 0$ for all $t \in [a,b]$, then the set S is linearly independent on the interval $[a,b]$.

First of all, let us acknowledge that the converse of the theorem is false. That is, if $\det(\mathbf{A}(t)) = 0$, we cannot conclude that the set S is linearly dependent. For example, the set of vectors

$$\left\{ \begin{pmatrix} t \\ 0 \end{pmatrix}, \begin{pmatrix} \cos t \\ 0 \end{pmatrix} \right\}$$

is linearly independent in $(-\infty, \infty)$, since it is impossible to find constants c_1 and c_2 such that $c_1 t + c_2 \cos t = 0$ for all $t \in (-\infty, \infty)$ (otherwise $\cos t = -\frac{c_1}{c_2} t$, a linear function). Nonetheless this set being linear independent, we still have

$$\begin{vmatrix} t & \cos t \\ 0 & 0 \end{vmatrix} = 0,$$

showing that the converse of 1.32 is false.

Another remark follows.

Remark

If a set S of vector functions is linearly dependent on an interval $[a, b]$, then it is also linearly independent at each point of $[a, b]$. In contrapositive form, this statement says that if S is linearly independent at every point of $[a, b]$, then it is linearly independent in $[a, b]$. However, the converse of this statement is false, that is, if S is linearly independent on $[a, b]$, we *cannot* conclude that S is linearly dependent at each point of $[a, b]$. For example,

$$S = \left\{ \begin{pmatrix} t \\ 0 \end{pmatrix}, \begin{pmatrix} \sqrt{t} \\ 0 \end{pmatrix} \right\}$$

is linearly independent in $[0, \infty)$, but at the point $t = 1$, we get the two vectors

$$\begin{pmatrix} 1 \\ 0 \end{pmatrix} \quad \text{and} \quad \begin{pmatrix} 1 \\ 0 \end{pmatrix},$$

and the set consisting of this two vectors is clearly linearly dependent.

Now we look at an example in which theorem 1.32 is applied.

Example 1.33

Consider the vectors $\mathbf{x}_1(t) = \begin{pmatrix} e^t \\ 1 \end{pmatrix}$ and $\mathbf{x}_2(t) = \begin{pmatrix} e^{-t} \\ 1 \end{pmatrix}$. Determine whether the set $S = \{\mathbf{x}_1(t), \mathbf{x}_2(t)\}$ is linearly dependent or linearly independent on $(-\infty, \infty)$.

Solution: We compute the determinant

$$\begin{vmatrix} e^t & e^{-t} \\ 1 & -1 \end{vmatrix} = -(e^t + e^{-t}),$$

and we see that this determinant is not zero for all values of t in $(-\infty, \infty)$. Hence, the set S is linearly independent.

1.2.10 Eigenvalues and Eigenvectors

Definition 1.34

An scalar λ is called an **eigenvalue** of an $n \times n$ matrix $\mathbf{A}$ and a nonzero vector $\mathbf{x}$ is an eigenvector associated with λ if and only if

$$\mathbf{Ax} = \lambda \mathbf{x}.$$

To derive an expression to find the eigenvalues, and their associated eigenvectors, of an $n \times n$ matrix $\mathbf{A}$, note that if $\mathbf{Ax} = \lambda \mathbf{x}$, then $\mathbf{Ax} - \lambda \mathbf{x} = \mathbf{0}$, where the left-hand side of this expression can be factored as $\mathbf{Ax} - \lambda \mathbf{x} = (A - \lambda \mathbf{I})\mathbf{x}$, and therefore we have

$$(\mathbf{A} - \lambda \mathbf{I})\mathbf{x} = \mathbf{0}.$$

By definition, $\mathbf{x}$ is a nonzero vector, so the matrix $(\mathbf{A} - \lambda \mathbf{I})$ must be nonsingular. This means that

$$\det(\mathbf{A} - \lambda \mathbf{I}) = 0. \tag{1.29}$$

Expanding the determinant on the left-hand side we obtain a polynomial in λ of degree n, which is called the **characteristic polynomial** of $\mathbf{A}$. Hence, **(1.29)** is a n-th degree polynomial equation in λ, and all of its n solutions $\lambda_1, \ldots, \lambda_n$ (some of which may be repeated) are eigenvalues of $\mathbf{A}$. The following example illustrates how to calculate the eigenvalues of a 2×2 matrix.

Example 1.35

Calculate the eigenvalues and eigenvectors of the matrix

$$\mathbf{A} = \begin{pmatrix} -2 & 1 \\ 1 & -2 \end{pmatrix}.$$

Solution: We have

$$\mathbf{A} - \lambda \mathbf{I} = \begin{pmatrix} -2 & 1 \\ 1 & -2 \end{pmatrix} - \begin{pmatrix} \lambda & 0 \\ 0 & \lambda \end{pmatrix} = \begin{pmatrix} -2 - \lambda & 1 \\ 1 & -2 - \lambda \end{pmatrix},$$

and then the characteristic polynomial of $\mathbf{A}$ is $\det(\mathbf{A} - \lambda \mathbf{I}) = (\lambda + 2)^2 - 1 = \lambda^2 + 4\lambda + 3 = (\lambda + 1)(\lambda + 3)$, so the equation **(1.29)** has, in this case, solutions $\lambda_1 = -1$ and $\lambda_2 = -3$, which are the two eigenvalues of the matrix $\mathbf{A}$. Now let us compute the eigenvector $\mathbf{x} = \begin{pmatrix} x_1 \\ x_2 \end{pmatrix}$ associated to $\lambda_1 = -1$. We solve the system

$$(\mathbf{A} - (-1)\mathbf{I})\mathbf{x} = \begin{pmatrix} -1 & 1 \\ 1 & -1 \end{pmatrix} \begin{pmatrix} x_1 \\ x_2 \end{pmatrix} = \mathbf{0},$$

and the solutions are $x_1 = x_2 = t$, where t is any number. Hence, any vector of the form

$$\mathbf{x_1} = \begin{pmatrix} t \\ t \end{pmatrix} = t \begin{pmatrix} 1 \\ 1 \end{pmatrix}$$

is an eigenvector associated to the eigenvalue $\lambda_1 = -1$. Using the same procedure as above, we find that any vector of the form

$$\mathbf{x_2} = t \begin{pmatrix} -1 \\ 1 \end{pmatrix}$$

is an eigenvector associated to the eigenvalue $\lambda_2 = -3$.

Let us now look at an example involving complex eigenvalues.

Example 1.36

Compute the eigenvalues and eigenvectors of the matrix

$$\mathbf{A} = \begin{pmatrix} -2 & 0 & 1 \\ 0 & 1 & -3 \\ -1 & 0 & -2 \end{pmatrix}.$$

Solution: In this case,

$$\mathbf{A} - \lambda \mathbf{I} = \begin{pmatrix} -\lambda - 2 & 0 & 1 \\ 0 & 1 - \lambda & -3 \\ -1 & 0 & -\lambda - 2 \end{pmatrix},$$

so

$$\begin{aligned} \det(\mathbf{A} - \lambda \mathbf{I}) &= (-2 - \lambda)(1 - \lambda)(-2 - \lambda) + 0 + 0 - (-1)(1 - \lambda)(1) - 0 - 0 \\ &= -\lambda^3 - 3\lambda^2 - \lambda + 5, \end{aligned}$$

and therefore the characteristic equation of matrix $\mathbf{A}$ can be written as

$$\lambda^3 + 3\lambda^2 + \lambda - 5 = 0.$$

We can find that $\lambda_1 = 1$ is a rational solution to the above equation. To find the other two solutions, we divide the polynomial $\lambda^3 + 3\lambda^2 + \lambda - 5$ by the factor $\lambda - 1$, which gives

$$\lambda^2 + 4\lambda + 5,$$

and therefore the other two solutions to the characteristic equation of $\mathbf{A}$ are the solutions to

$$\lambda^2 + 4\lambda + 5 = 0.$$

Such solutions are

$$\lambda_2 = -2 + i \qquad \text{and} \qquad \lambda_3 = -2 - i.$$

Hence, the eigenvalues of $\mathbf{A}$ are $\lambda_1 = 1$, $\lambda_2 = -2 + i$ and $\lambda_3 = -2 - i$. Let us now find the eigenvector $\mathbf{x_2}$ associated to the eigenvalue $\lambda_2 = -2 + i$. To do this, we solve the system

$$(\mathbf{A} - \lambda_2 \mathbf{I})\mathbf{x_2} = \mathbf{0}. \tag{1.30}$$

This can be done by performing row operations on the augmented matrix

$$\begin{pmatrix} -i & 0 & 1 & 0 \\ 0 & 3-i & -3 & 0 \\ -1 & 0 & -i & 0 \end{pmatrix}.$$

If we divide the first row of the above matrix by $-i$, we obtain

$$\begin{pmatrix} 1 & 0 & i & 0 \\ 0 & 3-i & -3 & 0 \\ -1 & 0 & -i & 0 \end{pmatrix}.$$

On this matrix, we now perform the row operation $R_3 \to R_1 + R_3$, which gives us

$$\begin{pmatrix} 1 & 0 & i & 0 \\ 0 & 3-i & -3 & 0 \\ 0 & 0 & 0 & 0 \end{pmatrix},$$

and then we perform $R_2 \to \left(\frac{1}{3-i}\right) R_2$ to obtain

$$\begin{pmatrix} 1 & 0 & i & 0 \\ 0 & 1 & -\dfrac{9}{10} - \dfrac{3}{10}i & 0 \\ 0 & 0 & 0 & 0 \end{pmatrix}.$$

This tells us that the solutions to system (1.30) are of the form $x_3 = t$, $x_2 = \left(\frac{9}{10} + \frac{3}{10}i\right)t$ and $x_1 = -it$, where t is any number. Hence, any vector of the form

$$\mathbf{x}_2 = t \begin{pmatrix} -i \\ \frac{9}{10} + \frac{3}{10}i \\ 1 \end{pmatrix},$$

where t is any nonzero number, is an eigenvector associated to the eigenvalue $\lambda_2 = -2 + i$. For example, for $t = 10$, we get the eigenvector

$$\begin{pmatrix} -10i \\ 9 + 3i \\ 10 \end{pmatrix},$$

which can also be written as the vector sum

$$\begin{pmatrix} 0 \\ 9 \\ 10 \end{pmatrix} + i \begin{pmatrix} -10 \\ 3 \\ 0 \end{pmatrix},$$

where the real and imaginary parts have been separated. We leave to the reader to verify that the associated eigenvectors for $\lambda_1 = 1$ and $\lambda_3 = -2 - i$ are, respectively, vectors of the form

$$\mathbf{x}_1 = t \begin{pmatrix} -1 \\ 0 \\ -i \end{pmatrix} \qquad \text{and} \qquad \mathbf{x}_3 = t \begin{pmatrix} i \\ \frac{9}{10} - \frac{3}{10}i \\ 1 \end{pmatrix}.$$

1.2.11   Diagonalization

We have mentioned already that if $\mathbf{D}$ is a diagonal matrix, then it is easy to compute $\mathbf{D}^k$, where k is a positive integer, for in this case the computation only requires rising each element of $\mathbf{D}$ to the k-th power. It turns out that sometimes we can factor a matrix $\mathbf{A}$ as the product $\mathbf{PDP}^{-1}$, where $\mathbf{P}$ is an invertible matrix and $\mathbf{D}$ is a diagonal matrix. If we can factor $\mathbf{A}$ into such form, we say that $\mathbf{A}$ is **diagonalizable**, and the process of finding the factorization $\mathbf{A} = \mathbf{PDP}^{-1}$ is called **diagonalizing** the matrix $\mathbf{A}$.

If we have diagonalized a matrix $\mathbf{A}$, then it is also easy to compute powers of it, because (as we ask the student to verify) we will have

$$\mathbf{A}^k = \mathbf{PD}^k\mathbf{P}^{-1}.$$

Now, the question is, when is a matrix diagonalizable? And if we know that a matrix is diagonalizable, how do we actually diagonalize it, that is, how do we compute the factorization $\mathbf{A} = \mathbf{PDP}^{-1}$? The following fact answers these questions.

Theorem 1.37

An $n \times n$ matrix $\mathbf{A}$ is diagonalizable if and only if it has n linearly independent eigenvectors. Moreover, if $\mathbf{A} = \mathbf{PDP}^{-1}$ with $\mathbf{D}$ a diagonal matrix, then P is a matrix whose columns are the eigenvectors of $\mathbf{A}$, and the entries in the diagonal of $\mathbf{D}$ are the associated eigenvalues.

Now we can look at an example on how to diagonalize a matrix that is diagonalizable.

Example 1.38

Determine if the following matrix is diagonalizable, and if it is so, diagonalize it and compute **5**.

$$\mathbf{A} = \begin{pmatrix} 1 & 0 & 4 \\ 7 & -1 & 5 \\ 9 & 0 & 1 \end{pmatrix}.$$

Solution: By theorem 1.37, we first have to find the eigenvectors of $\mathbf{A}$. The characteristic polynomial of $\mathbf{A}$ is

$$\det(\mathbf{A} - \lambda\mathbf{I}) = -\lambda^3 + \lambda^2 + 37\lambda + 35,$$

which has roots $\lambda_1 = 7$, $\lambda_2 = -5$ and $\lambda_3 = -1$. These are the eigenvalues of $\mathbf{A}$, and the associated eigenvectors are

$$\mathbf{x}_1 = \begin{pmatrix} 16 \\ 29 \\ 24 \end{pmatrix}, \quad \mathbf{x}_2 = \begin{pmatrix} -8 \\ -1 \\ 12 \end{pmatrix}, \quad \mathbf{x}_3 = \begin{pmatrix} 0 \\ 1 \\ 0 \end{pmatrix}.$$

(d) For what values of a will the function $y = x^a$ solve the DE

$$x^2 y'' = 2y.$$

(e) Given that $f(x)$ is continuous in an interval I, x_0 is in I and there exist an arbitrary y_0 satisfying $y(x_0) = y_0$, verify that the solution to the DE

$$\frac{dy}{dx} = f(x)$$

is given by

$$y(x) = y_0 + \int_{x_0}^{x} f(s) \, ds.$$

Is the solution $y(x)$ unique? Prove.

(3) Linear Algebra review problems.

(a) Use the definition of matrix exponentiation,

$$e^{t\mathbf{A}} = I + t\mathbf{A} + \frac{1}{2}t^2\mathbf{A}^2 + \cdots + \frac{1}{n!}t^n\mathbf{A}^n + \cdots$$

to compute $e^{t\mathbf{A}}$ where

$$\mathbf{A} = \begin{pmatrix} 0 & -1 \\ 1 & 0 \end{pmatrix}.$$

(b) Suppose $\mathbf{A} = \mathbf{P}^{-1}\mathbf{DP}$, where $\mathbf{A}$ and $\mathbf{P}$ are non singular $n \times n$ matrices and $\mathbf{D}$ is an $n \times n$ diagonal matrix. Verify

$$\mathbf{A}^n = \mathbf{P}^{-1}\mathbf{D}^n\mathbf{P}.$$

(c) Linear independence. Verify the following.
   i. $f(t) = \sin^t$ and $g(t) = 2 - 2\cos^2 t$ are linearly dependent.
   ii. $f(t) = t$ and $g(t) = t$ are linearly independent.
   iii. $f(t) = e^{2t}$ and $g(t) = e^{3t}$ are linearly independent.
   iv. $f(t) = \sin t$ and $g(t) = \cos t$ are linearly independent.

(d) Find the eigenvalues and eigenvectors of the following matrices. Also, compute the determinant of each matrix.
   i. $\begin{pmatrix} 6 & 16 \\ -1 & -4 \end{pmatrix}$
   ii. $\begin{pmatrix} 7 & -1 \\ 4 & 3 \end{pmatrix}$
   iii. $\begin{pmatrix} 5 & 0 \\ 0 & -1 \end{pmatrix}$
   iv. $\begin{pmatrix} 6 & 5 \\ -8 & -6 \end{pmatrix}$

$$\text{v.} \quad \begin{pmatrix} 1 & 2 & 5 \\ 0 & 3 & 8 \\ 0 & 0 & -4 \end{pmatrix}$$

(e) Important facts about eigenvalues and eigenvectors.

    i. Suppose that $\mathbf{A}$ is a matrix with all of its eigenvalues being real and distinct. Then the corresponding eigenvalues are linearly independent, otherwise check this fact against (d)-i and (d)-ii in the previous exercise.

    ii. The eigenvalues of a diagonal matrix are the diagonal entries of the matrix. Check this against (d)-iii.

    iii. Eigenvalues of a upper-triangular matrix are the diagonal entries of the matrix. Check this fact against (d)-v.

    iv. Suppose $\mathbf{A}$ is an $n \times n$ matrix with eigenvalues $\lambda_1, \lambda_2, \ldots, \lambda_n$. Then

$$\det(\mathbf{A}) = \lambda_1 \cdot \lambda_2 \cdots \lambda_n \qquad \text{the product of the eigenvalues,}$$

and
$$\text{trace}(\mathbf{A}) = \lambda_1 + \lambda_2 + \cdots + \lambda_n.$$

Check this against (d)-iii and (d)-iv.

Chapter 2

First Order ODEs

The generic form of a first order differential equation is

$$y' = f(x, y).$$

To study first order differential equations, we will distinguish two major aspects associated to them, namely their quantitative aspect and its qualitative aspect. Describing a DE quantitatively means actually finding an analytic expression that defines a solution to it, whereas a qualitative description of a differential equation will involve the discussion of its interval of validity, its existence and its uniqueness, and its relation to equilibrium solutions.

According to the techniques that we use to analyze a DE quantitatively, that is, to find expressions for its solutions, we can categorize the first order differential equations that we will study as indicated by the following diagram.

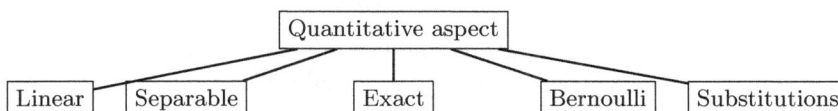

```
                    ┌─────────────────────┐
                    │ Quantitative aspect │
                    └─────────────────────┘
   ┌────────┐ ┌───────────┐   ┌───────┐   ┌───────────┐ ┌───────────────┐
   │ Linear │ │ Separable │   │ Exact │   │ Bernoulli │ │ Substitutions │
   └────────┘ └───────────┘   └───────┘   └───────────┘ └───────────────┘
```

The diagram above not only tells us about the subcategories into which we divide first order DE, but also helps us to separate the solution techniques that need to be discussed in each case.

On the other hand, we will categorize our qualitative study of first order differential equations as follows

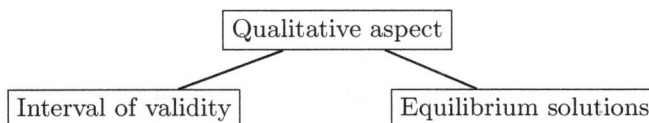

```
                    ┌────────────────────┐
                    │ Qualitative aspect │
                    └────────────────────┘
      ┌─────────────────────┐       ┌───────────────────────┐
      │ Interval of validity │       │ Equilibrium solutions │
      └─────────────────────┘       └───────────────────────┘
```

That is, in a qualitative description of the solutions to a first order DE we talk about the behaviour of solutions in terms of intervals of validity, existence and uniqueness of the solutions, and the relations of the solutions to possible equilibrium solutions.

2.1   Linear First Order DE

The generic form of **linear first order DE** is

$$y' + p(x)y = q(x), \qquad \text{where } p(x) \text{ and } q(x) \text{ are continuous.} \qquad (2.1)$$

Note that

(1) equation **(2.1)** is linear because y and y' are not mixed;

(2) the equation is of first order because y' is the highest derivative;

(3) $p(x)$ and $q(x)$ are continuous. Why? The necessity of this requirement will be seen shortly.

How do we solve equation (2.1)?   We need to find a factor, called the **integrating factor** for the given first order DE, which we will denote by $\mu(x)$. The integrating factor is given by

$$\mu(x) = e^{\int p(x)\ dx}. \qquad (2.2)$$

The integration involved in the computation of $\mu(x)$ explains why $p(x)$ must be continuous. Using the integrating factor $\mu(x)$, we can solve a first order DE by going through the steps of the following algorithm.

Algorithm to Solve Linear First Order DE

(1) If necessary, rewrite the DE to put it in the correct form, that is, in the form **(2.1)**.

(2) Find the integrating factor by using **(2.2)**.

(3) Multiply the DE by the integrating factor $\mu(x)$ and then note that the left-hand side becomes an exact differential: $(y(x) \cdot \mu(x))'$.

(4) Integrate both sides of the equation.

(5) Find the solution $y(x)$ by doing the necessary algebraic manipulations.

Let's explain more on item 3 of the algorithm. If we multiply **(2.1)** by the integrating factor $\mu(x)$, we get

$$\mu(x)\left[y'(x) + p(x)y(x)\right] = \mu(x)q(x).$$

We claim that the left-hand side is $(y(x)\mu(x))'$. To see this, let's expand

$$(y(x)\mu(x))' = y(x)\mu'(x) + \mu(x)y'(x).$$

Now, $\mu(x) = e^{\int p(x)\ dx}$, so

$$\mu'(x) = p(x)e^{\int p(x)\ dx}.$$

Therefore

$$
\begin{aligned}
y(x)\mu'(x) + \mu(x)y'(x) &= y(x)p(x)e^{\int p(x)\,dx} + e^{\int p(x)\,dx}y'(x) \\
&= e^{\int p(x)\,dx}\left[y(x)p(x) + y'(x)\right] \\
&= \mu(x)\left[y'(x) + y(x)p(x)\right],
\end{aligned}
$$

which is what we wanted to show.

Let's look at an example.

Example 2.1

Solve the differential equation

$$
\frac{dy}{dt} = 9.8 - 0.196y. \tag{2.3}
$$

Solution: Compare this equation to (2.1). We need to begin by writing equation (2.3) in the correct form:

$$
\frac{dy}{dt} + 0.196y = 9.8.
$$

Now we go to step (2), and find the integrating factor

$$
\mu(x) = e^{\int p(t)\,dt} = e^{\int 0.196\,dt} = e^{0.196t}.
$$

Now step (3). We multiply the DE (the one written in the correct form) by our integrating factor, and we note that the LHS becomes

$$
\left(y(t) \cdot e^{0.196t}\right)'.
$$

Therefore, the new DE is

$$
\left(y(t) \cdot e^{0.196t}\right)' = 9.8e^{0.196t}.
$$

Step (4) is to integrate both sides of this new DE:

$$
\begin{aligned}
\int \left(y(t) \cdot e^{0.196t}\right)'\,dt &= \int 9.8e^{0.196t}\,dt \\
\Rightarrow y(t) \cdot e^{0.196t} &= 9.8\frac{e^{0.196t}}{0.196} + C \\
\Rightarrow y(t) \cdot e^{0.196t} &= 50e^{0.196t} + C \\
\Rightarrow y(t) &= 50 + Ce^{-0.196t}.
\end{aligned}
$$

We have found our solution, $y(t) = 50 + Ce^{-0.196t}$. Let's not stop here, and do an analysis of this solution. First, note that what we have found is indeed a (one-parameter) family of solutions determined by the value of the constant of integration C. Let's ask *Mathematica* to plot a few integral curves corresponding to the values $C = -3, -2, -1, 0, 1, 2, 3$.

In[1]:= **Plot** $\left[\left(50 + \#1\, e^{-0.196\, t}\, \& \right) \right.$ **/@** $\{-3,\ -2,\ -1,\ 0,\ 1,\ 2,\ 3\}$, $\{$**t**$,\ -5,\ 10\}\left.\right]$

Out[1]:=

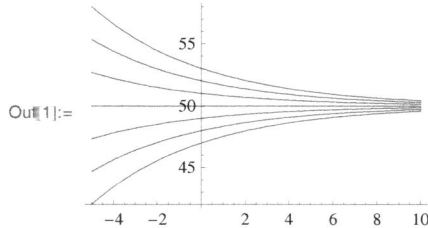

Note the following things:

(1) $y(t)$ is stable around the value $C = 0$.

(2) As long as C is finite, $y(t)$ is bounded, irrespective of the value of t.

Think about the following: what happens to the solution of example 2.1 if an initial condition is provided? For example, $y(0) = 96$.

Example 2.2

Solve the initial value problem

$$(\sin x)y' + (\cos x)y = 2\sin^3 x \cos x - 3, \qquad y(\pi/4) = \sqrt{2}, 0 < x < \pi/2. \quad \textbf{(2.4)}$$

Solution: First, we divide the DE by $\sin x$ to put it in the standard form **(2.1)**. This yields

$$y' + (\cot x)y = 2\sin^2 x \cos x - \frac{3}{\sin x}.$$

Next, we find

$$\mu(x) = e^{\int p(x)\ dx} = e^{\int \cot x\ dx} = e^{\ln(\sin x)} = \sin x.$$

We know that multiplying equation **(2.4)** by $\mu(x)$ makes the LHS the exact differential $(y(x) \cdot \mu(x))'$, whereas in the RHS would be $2\sin^3 x \cos x - 3$. Hence, the DE becomes

$$(y(x) \cdot \mu(x))' = 2\sin^3 x \cos x - 3.$$

If we integrate both sides of the above equation, we obtain

$$(y(x) \cdot \mu(x))' = \int 2\sin^3 x \cos x - 3\ dx$$

$$\Rightarrow y(x) \cdot \mu(x) = \frac{2\sin^4 x}{4} - 3x + C$$

$$y(x) \cdot \sin x = \frac{1}{2}\sin^4 x - 3x + C.$$

In this problem, we were given the initial condition $y(\pi/4) = \sqrt{2}$, that is, when $x = \pi/4$ we have $y = \sqrt{2}$. If we plug-in these values into the last equation obtained, we have

$$\sqrt{2}\frac{1}{\sqrt{2}} = \frac{1}{2}\left(\frac{1}{2}\right)^4 - 3\frac{\pi}{4} + C,$$

and if we solve for C we get

$$C = \frac{7}{8} + \frac{3\pi}{4}.$$

Hence,

$$y(x) \cdot \sin x = \frac{1}{2}\sin^4 x - 3x + \frac{7}{8} + \frac{3\pi}{4}$$
$$y(x) = \frac{1}{2}\sin^3 x - \frac{3x}{\sin x} + \frac{\frac{7}{8} + \frac{3\pi}{4}}{\sin x},$$

and this is the solution to our IVP problem.

Think The DE of this example is, in standard form

$$y' + (\cot x)y = 2\sin^2 x \cos x - \frac{3}{\sin x}.$$

Here, $p(x) = \cot x$. Is $p(x)$ continuous in the interval chosen $(0 < x < \pi/2)$? If not, how would you solve this DE?

Now we will use *Mathematica* to solve a linear first order DE. This can be done right away using the *Mathematica* function **DSolve**, but we will go through the solution process outlined by the algorithm for solving a linear first order DE. This will provide additional practice for you to learn the steps of the algorithm, and it will illustrate how to use *Mathematica* to check every computation you need to perform in the solution process.

Example 2.3

Using *Mathematica*, solve the differential equation

$$y' = \ln x + \frac{y}{x}.$$

Solution: First, we need to rewrite the differential equation:

$$y' - \frac{y}{x} = \ln x.$$

Let's give a name to our equation within *Mathematica*:

In[1]:= `Clear[eq]; eq = y'[x] -` $\frac{y[x]}{x}$ `== Log[x]`

Out[1]:= $-\frac{y[x]}{x}$ `+ y'[x] == Log[x]`

Comparing this equation with the form **(2.1)**, we see that, in this case, we have

$$p(x) = -\frac{1}{x}.$$

Let's use *Mathematica* to find the integrating factor:

In[2]:= `Clear[intFactor]; intFactor = e`$^{\int -\frac{1}{x}\,dx}$

Out[2]:= $\frac{1}{x}$

Now we multiply the equation **eq** by **intFactor**. To multiply both sides of the equation, we use the *Mathematica* function **Thread**:

In[3]:= `Clear[newEq]; newEq = Thread[intFactor eq, Equal]; Expand[newEq]`

Out[3]:= $-\frac{y[x]}{x^2} + \frac{y'[x]}{x} == \frac{Log[x]}{x}$

Now we simply integrate both sides of the equation (*Mathematica* will handle the integration of the exact differential on the left-hand side properly).

In[4]:= `Clear[sol]; sol = Thread[`$\int$`newEq dx, Equal]`

Out[4]:= $\frac{y[x]}{x} == \frac{Log[x]^2}{2}$

Mathematica does not add a constant of integration, so we have to include it ourselves:

In[5]:= `sol[[2]] = sol[[2]] + c; sol`

Out[5]:= $\frac{y[x]}{x} == c + \frac{Log[x]^2}{2}$

To obtain the final solution, let's ask *Mathematica* to isolate **y[x]**:

In[6]:= `Solve[sol, y[x]]`

Out[6]:= $\left\{\left\{y[x] \to \frac{1}{2}\, x\, \left(2\, c + Log[x]^2\right)\right\}\right\}$

Let's compare this result with the one given by the function **DSolve**.

In[7]:= `DSolve[eq, y[x], x]`

Out[7]:= $\left\{\left\{y[x] \to x\, C[1] + \frac{1}{2}\, x\, Log[x]^2\right\}\right\}$

This is the same answer we obtained. To complete the example, let's plot our solution for a few values of the constant of integration **c**:

2.1.1   Bernoulli Differential Equation

A **Bernoulli differential equation** is an DE of the form

$$y' + p(x)y = q(x)y^n, \tag{2.5}$$

where

(1) p and q are continuous functions in a given interval;

(2) $n \neq 0$ and $n \neq 1$. The reason for this is that $n = 0$ and $n = 1$ make the equation a linear first order DE, which is solved using an integrating factor(see section 2.1.)

Algorithm to Solve a Bernoulli DE

(1) Divide equation **(2.5)** by y^n to obtain

$$\frac{1}{y^n}y' + p(x)y^{1-n} = q(x).$$

(2) Use the substitution

$$v(x) = y^{n-1} \qquad \Rightarrow \qquad v'(x) = (1-n)y^{-n}y'.$$

(3) Then **(2.5)** becomes

$$\frac{1}{1-n}v'(x) + p(x)v(x) = q(x),$$

which is a linear first order DE and can therefore be solved using an integrating factor.

Example 2.4

Solve the following IVP.

$$y' + \frac{1}{x}y = xy^2, \qquad y(1) = 2,\ x > 0. \tag{2.6}$$

Solution: Since $n = 2$, this is a Bernoulli DE. We divide **(2.6)** to obtain

$$\frac{1}{y^2}y' + \frac{1}{x}\cdot\frac{1}{y} = x. \tag{2.7}$$

Use the substitution

$$v = y^{1-2} = y^{-1} \qquad \Rightarrow \qquad v' = -\frac{1}{y^2}y'.$$

We substitute these into **(2.7)**, which then changes to

$$-v' + \frac{1}{x}v = x,$$

that is,

$$v' - \frac{1}{x}v = -x.$$

This is a linear first order DE, which needs to be solved using an integrating factor:

$$\mu(x) = e^{\int -\frac{1}{x}\,dx} = e^{-\ln x} = e^{\ln \frac{1}{x}} = \frac{1}{x}.$$

Note that the above computation involved $\ln x$, which is only defined for $x > 0$. This explains the restriction made in **(2.6)** for the values of x.

Now we multiply our DE by $\mu(x) = \frac{1}{x}$:

$$\frac{1}{x}v' + \left(-\frac{1}{x^2}\right)v = -1,$$

that is,

$$\left(\frac{1}{x}\cdot v\right)' = -1.$$

Integrating both sides we get

$$\frac{1}{x}\cdot v = -x + C.$$

Now recall that $v = \frac{1}{y}$, so we can now write

$$\frac{1}{x}\cdot\frac{1}{y} = -x + C.$$

To find the value of C, use the initial condition $y(1) = 2$:

$$\frac{1}{1}\cdot\frac{1}{2} = -1 + C,$$

so $C = \frac{3}{2}$. Hence

$$\frac{1}{x}\cdot\frac{1}{y} = -x + \frac{3}{2}.$$

Solving the above expression for y gives

$$y = \frac{2}{x(3-2x)}.$$

Let's now analyze this solution quickly. For $x = 0$ and $x = \frac{3}{2}$, the solution blows up to infinity. If $0 < x < 3/2$ or $3/2 < x < \infty$, the solution will remain finite.

2.2   Separable Differential Equations

A differential equation is **separable** if it can be written in the form

$$F(y)\frac{dy}{dx} = G(x).$$

Function of y Function of x

Therefore, in a separable differential equation, we can write one side of the equation as the product of $F(y)$ and the derivative $\frac{dy}{dx}$, and such that the other side contains $G(x)$ only.

For a given separable differential equation, we may not get an explicit solution, that is, a solution given by an expression $y = f(x)$, but instead we can get an implicit solution given by an expression $f(x, y) = 0$.

Let's look at some examples.

Example 2.5

Solve the IVP

$$\frac{dy}{dx} = 4y^3 x, \quad y(1) = \frac{1}{64}.$$

Solution: If the above equation is separable, we should be able to put it in the form

$$F(y)\frac{dy}{dx} = G(x).$$

To do this, we divide the DE by y^3, which yields

$$\frac{1}{y^3}\frac{dy}{dx} = 4x.$$

This is now in the correct form of a separable DE. We now integrate both sides of the equation to obtain

$$\int \frac{1}{y^3}\frac{dy}{dx} = \int 4x$$
$$-\frac{1}{2y^2} = 4\frac{x^2}{2} + C$$
$$-\frac{1}{2y^2} = 2x^2 + C.$$

Using the initial condition $y(1) = \frac{1}{64}$, we find that

$$-\frac{1}{2 \cdot \left(\frac{1}{64}\right)^2} = 2 \cdot 1^2 + C,$$

so $C = -2050$. Therefore,

$$-\frac{1}{2y^2} = 2x^2 - 2050.$$

Now we solve for y to find the explicit solution:

$$y = \pm\frac{1}{-4x^2 + 4100}$$

Question: We said that a first order differential equation with an initial condition should have a unique solution, so what is wrong in this case? Apparently we get two solutions because of the $\pm$ sign in our result. Let's check, using the initial condition, if it is true that either of these signs will yield a solution. If we plug-in the value $x = 1$, we get

$$\frac{1}{-4x^2 + 4100} = \frac{1}{64} \qquad \text{and} \qquad -\frac{1}{-4x^2 + 4100} = -\frac{1}{64}.$$

We see that only the first expression gives the correct value for y indicated by the initial condition, so the solution to our IVP is

$$y = \frac{1}{-4x^2 + 4100}.$$

This solution makes sense as long as $4100 - 4x^2 > 0$, that is, as long as $x^2 < 1025$. Using approximate values, this means that

$$-32.01 < x < 32.01,$$

which is the interval of validity of the solution to our IVP. Finally, we ask *Mathematica* to plot this solution:

In[1]:= $\texttt{Plot}\left[\frac{1}{-4\,x^2+4100}\,,\ \{x,\,-33,\,33\},\ \texttt{Exclusions} \rightarrow \left\{-4\,x^2 + 4100 == 0\right\}\right]$

Out[1]:=

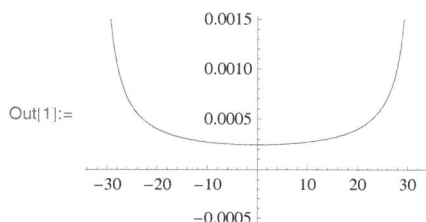

Example 2.6

Solve the IVP

$$\frac{dy}{dx} = \frac{xy^2}{\sqrt{1 + x^2}}, \qquad y(0) = 1.$$

Solution: To separate the variables in this DE, we divide by y^2, and then integrate both sides of the equation:

$$\begin{aligned}
\frac{1}{y^2}\frac{dy}{dx} &= \frac{x}{\sqrt{1 + x^2}} \\
\int \frac{1}{y^2}\frac{dy}{dx} &= \int \frac{x}{\sqrt{1 + x^2}} \\
-\frac{1}{y} &= \sqrt{1 + x^2} + C.
\end{aligned}$$

Now we use the initial condition $y(0) = 1$ to find that

$$-\frac{1}{1} = \sqrt{1} + C \qquad \Rightarrow \qquad C = -2.$$

Hence,

$$-\frac{1}{y} = \sqrt{1 + x^2} - 2,$$

and then

$$y = \frac{1}{2 - \sqrt{1 + x^2}}.$$

Note that as $x \to \pm\infty$, $y \to 0$, and that the solution y is not defined for $x^2 = 3$.

Finally, we plot the function using *Mathematica*:

In[1]:= **Plot** $\left[\dfrac{1}{2-\sqrt{1+x^2}} \textbf{, \{x, -5, 5\}, Exclusions} \rightarrow \left\{ \textbf{x}^2 \texttt{==} \textbf{3} \right\} \right]$

Out[1]:=

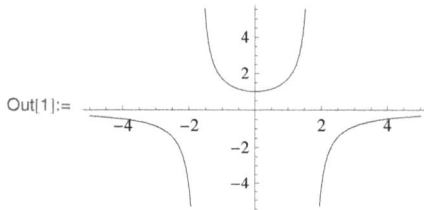

2.3 Exact Differential Equations

The formal structure we will look at this section is

$$F(x, y) + G(x, y)\frac{dy}{dx} = 0. \tag{2.8}$$

Note the following things about this form of differential equation:

(1) One side of the equation must be equal to zero.

(2) There must be a "+" between the two terms on the other side of the equation.

(3) Hopefully, there exists a "magic function", say $\varphi(x, y)$, such that

$$\varphi_x = F(x, y) \qquad \text{and} \qquad \varphi_y = G(x, y).$$

Conditions (1)–(3) above together will make **(2.8)** an **exact differential equation**.

Test for Exactness   Note that if the function φ mentioned above is continuous, and so are $\varphi_x = F$ and $\varpi_y = G$, then there is a relation between the second mixed partial derivatives $\varphi xy = F_y$ and $\varphi_{yx} = G_x$. What is this relation? The relation is

$$\varphi xy = \varphi_{yx} \qquad \Rightarrow \qquad F_y = G_x$$

In fact, if this equality is true, that is, if

$$F_y = G_x,$$

then the differential equation **(2.8)** is exact, and this is how we check for exactness.

How to solve equation (2.8) if it is exact?   By condition (3), equation **(2.8)** can be written as

$$\begin{aligned} \varphi_x + \varphi_y \frac{dy}{dx} &= 0 \\ \Rightarrow \qquad \varphi_x \, dx + \varphi_y \, dy &= 0 \\ \Rightarrow \qquad \frac{d}{dx}\varphi(x,y) &= 0. \end{aligned}$$

Integrating both sides,

$$\varphi(x,y) = C,$$

which is an expression that gives us an implicit solution to the differential equation **(2.8)**.

Example 2.7

Determine if the differential equation below is exact. If so, find a solution satisfying the given initial condition.

$$2xy - 9x^2 + (2y^2 + x^2 + 1)\frac{dy}{dx} = 0, \qquad y(0) = 1.$$

Solution: The equation is in the standard form **(2.8)** that we need, with

$$F(x,y) = 2xy - 9x^2 \qquad \text{and} \qquad G(x) = 2y^2 + x^2 + 1.$$

From this we obtain

$$F_y = 2x \qquad \text{and} \qquad G_x = 2x,$$

and therefore $F_y = G_x$ and the DE is exact. Next, we use the fact that

$$\varphi_x = F.$$

Integrating both sides, we have

$$\int \varphi_x(x, y) \; dx \;\; = \;\; \int F(x, y) \; dx$$

$$\varphi \;\; = \;\; \int (2xy - 9x^2) \; dx$$

$$\varphi \;\; = \;\; yx^2 - \frac{3x^3}{3} + h(y). \qquad \textbf{(2.9)}$$

Note that the constant of integration is a function h of y. Why do we have such constant of integration? Because any function of y alone disappears after differentiation with respect to x, so to go from φ_x to φ we need to recover this function.

Now, how do we find the function $h(y)$? If we differentiate both sides of $\varphi = yx^2 - \frac{3x^3}{3} + h(y)$ with respect to y, we have

$$\varphi_y = x^2 + h'(y),$$

but recall that by definition $\varphi_y = G$, where $G = 2y^2 + x^2 + 1$. Therefore,

$$x^2 + h'(y) \;\; = \;\; 2y^2 + x^2 + 1$$
$$h'(y) \;\; = \;\; 2y^2 + 1,$$

and integrating both sides we have

$$h(y) = \frac{2}{3}y^3 + y + C.$$

Now we plug-in this function into **(2.9)** to get our solution:

$$\varphi(x, y) = yx^2 - \frac{3x^3}{3} + \frac{2}{3}y^3 + y + C.$$

Disclaimer: It is easy to get confused between the method of separation of variables and the method of solving exact differential equations. For example, consider

$$\frac{dy}{dx} = y^2 x \qquad \text{and} \qquad 2xy - 9x^2 + (2y^2 + x^2 + 1) = 0.$$

The first equation can also be written as

$$\frac{dy}{dx} + (-y^2 x) = 0.$$

But this equation is separable, and can be solved by the method explained in section 2.2. Why don't we use the same method to solve $2xy - 9x^2 + (2y^2 + x^2 + 1) = 0$? The reason is just that the variables x and y can't be separated in this equation, so we have to try another method. Since the equation is exact, we can solve it as we did in the last example.

Example 2.8

Solve the following IVP,

$$2xy^2 + 4 = 2(3 - x^2 y)y', \qquad y(1) = 4. \qquad (2.10)$$

Solution: Separation of variables will not work in this case. Why? Again, this is simply that we cannot separate x and y, and we need to resort to another method. How about this equation being exact and therefore solvable by the method used in the last example? Let's bring the equation into the standard form **(2.8)**.

$$2xy^2 + 4 - 2(3 - x^2 y)y' = 0$$
$$\Rightarrow \qquad \underbrace{2xy^2 + 4}_{F(x,y)} + \underbrace{2(x^2 y - 3)}_{G(x,y)} y' = 0.$$

Now we find that

$$F_y = 2x \cdot 2y = 4x$$
$$G_x = 4xy,$$

so equation **(2.10)** is exact. How, then, do we solve the DE? Being that this equation is exact, there exists a function $\varphi(x, y)$ such that

$$\varphi(x, y) = \int G(x, y)\, dy \qquad \text{or} \qquad \varphi(x, y) = \int F(x, y)\, dx.$$

Both of the above relations give us the function φ. Let's use

$$\varphi(x, y) = \int G(x, y)\, dy.$$

Then

$$\begin{aligned}
\varphi(x, y) &= \int G(x, y)\, dy \\
&= \int 2(x^2 y - 3)\, dy \\
&= 2\left(\int x^2 y\, dy - \int 3\, dy \right) \\
&= \left(\frac{x^2 y^2}{2} - 3y \right) + \begin{matrix} \nearrow h(x) \\ \searrow h(y) \end{matrix}
\end{aligned}$$

The lines in the last equation point to two choices of a constant of integration, one a function of x and one a function of y, but only one of the choices is correct. Since we are integrating with respect to y, the constant of integration must be a function of x, which it is lost after computing φ_y, and we need to recover it. Hence,

$$\varphi(x, y) = x^2 y^2 - 6y + h(x). \qquad (2.11)$$

We know $\varphi_x = F$ and $\varphi_y = G$, so from (2.11), we have

$$G = \varphi_x = 2xy^2 + h'(x),$$

but $G = 2xy^2 + 4$, so

$$2xy^2 + h'(x) = 2xy^2 + 4,$$

which means that $h'(x) = 4$. Integrating both sides, we find that

$$h(x) = 4x + C_1.$$

Substituting this into (2.11), we have

$$\varphi(x, y) = x^2 y^2 - 6y + 4x + C_1.$$

As discussed, the solution to an exact differential equation is always given by the expression

$$\varphi(x, y) = C,$$

so

$$x^2 y^2 - 6y + 4x + C_1 = C.$$

Recall that we usually group our constants of integration and make it a single constant, so if we define $K = C - C_1$ we can write

$$x^2 y^2 - 6y + 4x = K.$$

But know we have to use the initial condition given with the DE to find the value of K. Since $y(1) = 4$, we have

$$4^2 - 6 \cdot 4 + 4 \cdot 1 = K,$$

so $K = -4$. Finally, we have our implicit solution:

$$x^2 y^2 - 6y + 4x = -4.$$

Let's now write a list of steps that summarizes the process we have seen in the two last examples to solve an exact DE.

Algorithm to Solve an Exact DE

(1) Write in the standard form (2.8) and check for exactness:

   (a) The right-hand side must be equal to zero.

(b) The sign separating the two terms F and G must be a +.

(c) Write the equation in the form

$$F(x, y) + G(x, y) \frac{dy}{dx} = 0.$$

(d) Check if $F_y = G_x$.

(e) If we have the above equality, then the implicit solution to the DE is $\varphi(x, y) = C$, where $\varphi_x = F$ and $\varphi_y = G$.

(2) Now the solution process:

(a) We have

$$\varphi(x, y) = \int F(x, y) \, dx \qquad \text{and} \qquad \varphi(x, y) = \int G(x, y) \, dy,$$

and any of the above relations can be used to find φ.

(b) Suppose we use $\varphi(x, y) = \int F(x, y) \, dx$. Then the result of this integration must include a constant of integration $h(y)$ (a function of y). If the other integral, $\varphi(x, y) = \int G(x, y) \, dy$, is used, then the integration includes a constant of integration $h(x)$ (a function of x).

(c) If we use $\varphi(x, y) = \int F(x, y) \, dx$, we need to find the constant of integration $h(y)$. How? We differentiate $\varphi(x, y) = \int F(x, y) \, dx$ with respect to y, and use the fact that $\varphi_y = G$.

(d) The step above will give an expression for $h'(y)$. Upon an integration, $h(y)$ is obtained.

(e) Finally, we can obtain $\varphi(x, y) = C$ as our implicit solution. If any initial conditions are given, apply them to find the value of C.

2.4   Homogeneous Differential Equations

The term homogeneous certainly makes reference to the notion of "two things being equal" in some sense. Here, the homogeneity is related to the degree of the terms of an expression. More exactly, an expression is homogeneous if all of its terms have the same degree.

Are the following homogeneous?

(1) $xyy' + x^2 + y = 0$ $\Rightarrow$ No, the first and second terms are of second degree, while the last term is of first degree.

(2) $xyy' + x^2 + y^2 = 0$ $\Rightarrow$ Yes, all terms are of second degree

(3) $x^2yy' + x^3 + y^2 = 0$ $\Rightarrow$ No, not all terms are of the same degree.

(4) $y' + \sin(y/x) = 0$ $\Rightarrow$ Yes, all terms are of zero degree.

(5) $y' + \sin(xy) = 0$ $\Rightarrow$ No, the second term is of second degree.

Algorithm to Solve a Homogeneous Differential Equation

(1) Check that the DE is homogeneous.

(2) Use a substitution $y = vx$, where v is a function of x.

(3) Then $y' = v + xv'$.

(4) Apply the substitution and do some algebra to write the DE in terms of v and x.

(5) Solve the resulting differential equation.

Note: In the class of homogeneous differential equations, x will never take on the value zero.

Example 2.9

Solve the following IVP.

$$xyy' + x^2 + y^2 = 0, \qquad y(1) = 4, \ x > 0.$$

Solution: First, we see that this is a homogeneous equation. Next, we do the substitution

$$y = vx \qquad \Rightarrow \qquad y' = v + xv'.$$

Substituting these into the original DE, we get

$$x(vx)(v + xv') + x^2 + v^2x^2 = 0.$$

Dividing by x^2 (which is possible, since we said that when considering homogeneous DE x is never taken to be zero),

$$
\begin{aligned}
v(v + xv') + 1 + v^2 &= 0 \\
v(v + xv') &= -(1 + v^2) \\
v + xv' &= \frac{1 + v^2}{v} \\
xv' &= -\frac{1 + v^2}{v} - v = -\left(\frac{1 + 2v^2}{v^2}\right) \\
\frac{v}{1 + 2v^2}\frac{dy}{dx} &= -\frac{1}{x} \\
\int \frac{v}{1 + 2v^2}\, dv &= \int -\frac{1}{x}\, dx \\
\frac{1}{4}\ln(1 + 2v^2) &= -\ln x + \ln C \\
\frac{1}{4}\ln(1 + 2v^2) &= \ln\frac{C}{x} \\
(1 + 2v^2)^{1/4} &= \frac{C}{x} \\
1 + 2v^2 &= \left(\frac{C}{x}\right)^4.
\end{aligned}
$$

Since $y = vx$, $v = \frac{y}{x}$, so

$$
1 + 2\frac{y^2}{x^2} = \frac{C^4}{x^4},
$$

that is

$$
x^4 + 2x^2 y^2 = C^4.
$$

Applying the initial condition, we find the value of C to be $\sqrt[4]{33}$. Hence

$$
x^4 + 2x^2 y^2 = 33.
$$

We leave it to the student to solve the above expression for y and do a quick analysis of the solution.

Example 2.10

Solve the differential equation

$$
\frac{dy}{dx} = -2x + 2y2x + y. \tag{2.12}
$$

Solution: Let's rewrite the DE (2.12) as

$$
(2x + y)\frac{dy}{dx} + 2x - 5y = 0
$$

Both terms terms $(2x + 1)$ and $(2x - 5y)$ are of degree 1, so the DE is homogeneous.

Comment: It is also easy to check if a certain DE is homogeneous in the form given in (2.12). In this case, the degree of the LHS is 0, and the degree of the RHS is obtained by subtracting the degree of the denominator from that of the denominator, so the degree is also $1 - 1 = 0$.

Since (2.12) is homogeneous, we use the substitution

$$y = vx \qquad \Rightarrow \qquad y' = v + xv'.$$

Then (2.12) becomes

$$v + x\frac{dv}{dx} = \frac{-2x + 5vx}{2x + vx}$$

$$v + x\frac{dv}{dx} = \frac{-2 + 5v}{2 + v}$$

$$x\frac{dv}{dx} = \frac{-2 + 5v}{2 + v} - v,$$

which can be written as

$$\frac{2 + v}{-2 + 3v - v^2} \, dv = \frac{1}{x} \, dx.$$

Integrating both sides,

$$-\int \frac{2 + v}{-2 + 3v - v^2} \, dv = \int \frac{1}{x} \, dx$$

$$-\int \frac{3}{v - 1} - \frac{4}{v - 2} \, dv = \ln x + \ln C$$

$$-3\ln(v - 1) + 4\ln(v - 2) = \ln x + \ln C$$

$$\frac{(v - 2)^4}{(v - 1)^3} = Cx.$$

Using the substitution $v = y/x$ and simplifying a little bit, the last equation above becomes

$$-\frac{x(x - y)^3}{7(y - 2x)^4} = Cx.$$

This expression implicitly defines a one-parameter family of solutions to our DE.

2.5 Existence and Uniqueness

In this section we will turn our attention to a little bit of theory on first order differential equations, with the intention of trying to figure out if a solution to a DE actually exists and if this solution is unique, this without solving the DE whenever possible.

Note: Not all first order DE are solvable by using analytical techniques. In fact, there are many differential equations that cannot be solved analytically. Therefore, it is all the more important to look at the theory of existence and uniqueness.

Remark

Consider the following differential equations.

$$\begin{aligned} y' &= y, \\ y' &= \sin(y). \end{aligned}$$

Is there an obvious solution to these differential equations? Yes, the function $y(x) = 0$, which is called the trivial solution of these differential equations.

For our discussion of the uniqueness of a solution, trivial solutions will never be considered, meaning, we will always look at nontrivial (nonzero) solutions.

We will discuss two theorems of existence and uniqueness:

$$\boxed{\text{Existence and uniqueness}}$$

First theorem: applies to linear first order differential equations, that is, to equations of the form

$$y' + p(x)y = q(x),$$

with an initial condition $y(x_0) = y_0$.

Second theorem: applies also to nonlinear first order differential equations, that is, to equations of the form

$$y' = f(x, y),$$

with an initial condition $y(x_0) = y_0$.

Let's look at the first theorem of uniqueness and existence.

Theorem 2.11
Consider the DE

$$y' + P(x)y = q(x), \qquad y(x_0) = y_0, \tag{2.13}$$

where the following conditions are met:

(1) *$p(t)$ and $q(t)$ are continuous functions in the interval (a, b);*

(2) *the initial value x_0 is contained in this interval (a, b).*

Then there exists *a* unique *solution to the differential equation* **(2.13)**.

Note: The above theorem also helps us find the interval of validity of the solution to a DE.

Example 2.12

Determine if the IVP below has a unique solution, and in which interval this is so.

$$(t^2 - 1)y' + y = \ln|20 - 5t|, \qquad y(4) = 10.$$

Solution: Divide the DE by $t^2 - 1$ to obtain

$$y' + \frac{1}{t^2 - 1}y = \frac{\ln|20 - 5t|}{t^2 - 1},$$

which is a linear first order DE in the standard form **(2.1)**. In order for us to use theorem 2.11, we need continuity issues of

$$p(t) = \frac{1}{t^2 - 1} \qquad \text{and} \qquad q(t) = \frac{\ln|20 - 5t|}{t^2 - 2}$$

The function $p(t)$ has discontinuities at $t = \pm 1$, and $q(t)$ has discontinuities at $t = \pm 1$ and $t = 4$. Combining intervals of continuity of $p(t)$ and $q(t)$ we get

$$(-\infty, -1), \quad (-1, 1) \quad (1, 4) \quad (4, \infty).$$

Now let's look at the second condition of theorem 2.11. We need $t_0 = 4$ to be contained in one of these intervals. However, none of the intervals contains $t_0 = 4$, so we cannot use the theorem to conclude the existence and the uniqueness of the solution to our IVP. Can we change the initial condition to have a unique solution? Yes, we can take a different t_0, as for example $t_0 = 5$, which will be contained in the interval $(4, \infty)$ above.

We will now look at the second theorem of existence and uniqueness.

Theorem 2.13

Consider the IVP

$$y' = f(x, y), \qquad y(x_0) = y_0. \tag{2.14}$$

Suppose the following conditions are met:

(1) $f(x, y)$ *and* $\partial f/\partial y$ *are continuous in a rectangle* $a < x < b$ *and* $c < x < d$;

(2) *the rectangle above contains the point* (x_0, y_0).

Then there exists a unique solution to the IVP **(2.14)** *in some interval* $(x_0 - h, x_0 + h)$ *which is contained in* (a, b).

Note: There are some things we want to say about this theorem:

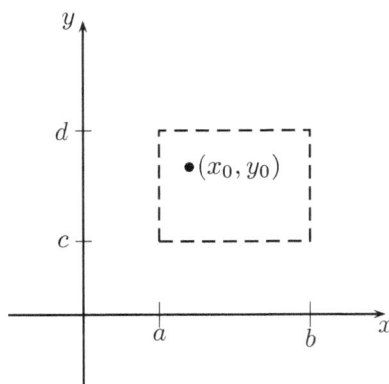

Figure 2.1.

(1) The existence and uniqueness phenomenon is not only at the point (x_0, y_0), but could be stretched by $(x_0 - h, x_0 + h)$ as long as it is within the rectangular boundary $(a, b) \times (c, d)$ (see figure 2.1.)

(2) This theorem doesn't say anything about the interval of validity, unlike the first theorem of existence and uniqueness.

(3) The above remark implies that for us to find the interval of validity of a solution, we have to actually solve the DE.

(4) Unlike the case in the first theorem, the interval of validity depends both on x_0 and y_0.

Example 2.14

Determine if the following IVP has a unique solution.

$$y' = y^{1/3}, \qquad y(1) = 0.$$

Solution: Here, $f(x, y) = y^{1/3}$, and then

$$\frac{\partial f}{\partial y} = \frac{1}{3} y^{-2/3} = \frac{1}{3y^{2/3}},$$

which is not continuous at $y = 0$. Hence, f and $\partial f / \partial y$ are both continuous on the rectangle above or below the x axis, without including this axis. But any of these rectangles doesn't contain $(1, 0)$, so there is not a unique solution.

$\boxed{\textbf{Think}}$ Is there a unique solution if the initial condition were $y(1) = 2$?. Yes, there is a unique solution, since $f(x, y)$ and $\partial f / \partial y$ are both continuous in the rectangle above the x-axis, which contains the point $(x_0, y_0) = (1, 2)$. Hence, by theorem 2.13, there exists a unique solution to the IVP.

In the example above, we have seen an IVP that doesn't have a unique solution. The next question is, then, if an IVP has no unique solution, can we find all the solutions? Let's look at an example.

Example 2.15

Solve the IVP

$$y' = y^{1/3}, \qquad y(0) = 0.$$

Solution: To begin with, we know that this DE has a trivial solution $y(x) = 0$. Now, this DE is separable, and it can be written as

$$\frac{dy}{y^{1/3}} = dt,$$

so integrating both sides yields

$$\frac{3}{2}y^{2/3} = t + C.$$

Applying the initial condition, we have

$$\frac{3}{2}0^{2/3} = 0 + C,$$

so $C = 0$. Hence,

$$y^{2/3} = \frac{2}{3}t \qquad \Rightarrow \qquad y = \pm\left(\frac{2}{3}t\right)^{3/2}.$$

Therefore, we have two possible solutions to our IVP because of the $\pm$ sign.

Remark
Without the knowledge of the existence and uniqueness theorem 2.13, we would have to go through the process of solution to figure out whether there exists a unique solution or not.

Exercises

Note: Below is a collection of ODEs for you to classify. You may not need to solve those completely, but what you must do is identify each ODE as exact, linear, Bernoulli, etc. and then comment on the solution technique. Let's proceed!

(1) $xy' = y + x$, $y(1) = 0$.

(2) $2x^2y\,dx + (1 + x^3)\,dy = 0$.

(3) $y\,dx + x\,dy = 0$.

(4) $y\,dx - x\,dy = 0$.

(5) $y' - \frac{4}{x}y = x^4$, $y(2) = k$, where $k \neq 0$.

(6) $y' + xy = xy^3$.

Summary of first order DEs:

Method	Standard form
Separation of variables	$F(y)\dfrac{dy}{dx} = G(x)$
Linear	$y' + p(x)y = q(x)$
Bernoulli	$y' + p(x)y = q(x)y^n$
Homogeneous	The terms in the equation, except the y', are of the same degree
Exact	$F(x,y) + G(x,y)\dfrac{dy}{dx}$

(1) Solve the following DE with the appropriate method. Double check the nature of the DE.

(a) $y' + \dfrac{1}{x}y = 3x - 2$.

(b) $y' + y = \sin x$.

(c) $(y + 1)\sin x\,dx + dy = 0$.

(d) $xy' + y = x^2 + 1$.

(e) $y' + xy = xy^2$.

(f) $y' = \dfrac{e^{x+y}}{e^{y-x}}$.

(g) $y' = \dfrac{\sin(x + y)}{\cos(x + y)}$.

(h) $(1 + y^3)\,dx + (3xy^2 + y + 1)\,dy = 0$.

(i) $(x^2 + \sin y)\dfrac{dy}{dx} = 2xy$.

(j) $(x + y)\,dx - x\,dy = 0$.

(k) $(x^2 + y)\,dx - y^2\,dy = 0$.

(l) $y' = x\sqrt{1 - y^2}$.

(2) Some applications problems.

(a) An object of mass m is dropped from a moving aircraft. The equation of motion of the object is given by the DE

$$\frac{dv}{dt} + \frac{Kv}{m} = g.$$

Assuming that $v = 0$ when $t = 0$ (why?), show, by solving the DE, that the velocity v at time t is given by

$$v(t) = \frac{mg}{K}\left(1 - e^{Kt/m}\right).$$

Analyze the solution, especially when $t \to \infty$. How do you explain this physically?

(b) The current flow in a simple electric circuit is governed by the DE

$$L\frac{dI}{dt} + RI = E_0,$$

where E_0 is constant (called the forcing function), I is the current, R the resistance, and L is the inductance.

   i. Solve the DE by the appropriate method.

   ii. Now solve the DE when the forcing function is $E_0\sin(\omega t)$.

   iii. Rewrite the solution obtained in ii in the following form

$$I = ce^{-(R/L)t} + \frac{E_0}{\sqrt{R^2 + \omega^2 L^2}}\sin\left(\omega t + \tan^{-1}\left(-\frac{\omega L}{T}\right)\right).$$

What happens when $t \to \infty$.

(c) Financial model: A corporation wishes to re-invest part of its earnings following a model given by

$$\frac{dY}{dt} = rY + P, \qquad Y(0) = 0,$$

where P is the principal sum, r the rate of interest compounded continuously, and Y the amount of growth.

   i. Solve the DE for $Y(t)$.

   ii. If $P = \$75,000$, $r = 8\%$ and $t = 10$ years, find $Y(t)$.

   iii. Find the time period, t, if $P = \$50,000$, $r = 9.25\%$ and $Y = \$650,000$.

(d) Decay Problem: A toxic asset in a major investment firm decays following the model

$$\frac{dy}{dt} = -kA,$$

where k is a constant that augments the decay further at any given time. Solve the DE for $A(t)$ and comment on the nature of the decay of the asset.

(3) **Exploratory Exercises**

(a) Consider the IVP

$$y' = x^3\sin^2 y - y\ln x, \qquad y(1) = 4.$$

Also, $1 \le x \le 2$ and $0 \le y \le 3$. Does the above have a unique solution according to what we understood from the discussion on uniqueness?

(b) Consider the IVP
$$y' = ky, \qquad y(0) = 1,$$
along with the solution
$$y(x) = 1 + \int_0^x ky'(t)\, dt.$$

Is the solution unique in the neighborhood of $x = 0$? (Hint: Begin by contradicting the conjecture, i.e. assume there exist two solutions $y_1(x)$ and $y_2(x)$. Then try to show that $y_1(x) = y_2(x)$.)

(c) Consider the IVP
$$y' = 3y^{2/3}, \qquad y(0) = 0.$$

  i. Show that $y(x) = x^3$ is a solution to the IVP.
  ii. Argue that $y(x) = 0$ is also a solution.
  iii. Does the IVP have a unique solution?

(d) Consider the IVP
$$x' = |x|, \qquad x(0) = 0.$$

Comment on the solution of this IVP in terms of existence and uniqueness.

(e) Some higher order DE can be solved by a technique called **reduction of order** (if the dependent or the independent variable is missing). For example, the IVP
$$y'' = y'e^{2y}, \qquad y(0) = 0,\ y'(0) = 1,$$
could be solved by this method: Set
$$y' = v,$$
so $y'' = v\frac{dv}{dy}$ where v is a function of y, and the equation becomes
$$v\frac{dv}{dy} = ve^y.$$

Finish the problem and find $y(x)$.

(f) Use the above technique to solve the DE
$$2yy'' = 1 + (y')^2$$

(g) A very basic oscillation motion is governed by
$$y'' + k^2 y = 0, \qquad \text{where } k \in \mathbb{R}.$$

Use the method of reduction of order (setting $y' = v$ so that $y'' = v\,dv/dy$) to solve the DE and verify that the solution is
$$y(x) = \sqrt{A}\sin(\pm Bx + C),$$
with A, B, C being arbitrary constants of integration.

Chapter 3

Second Order Linear Differential Equations

In this chapter we will study second order linear DE, that is, equations of the form

$$a(x)y'' + b(x)y' + c(x)y = g(x). \tag{3.1}$$

Here are some quick facts about this kind of DE:

(1) The equation includes second order derivatives.

(2) We need to integrate twice to get the solution.

(3) The solution will include two constants of integration.

Our discussion of these differential equations will be mostly restricted to second order linear differential equations with constant coefficients, that is, differential equations like **(3.1)** for which

$$\left. \begin{array}{l} a(x) = a \\ b(x) = b \\ c(x) = c \end{array} \right\} \text{ are all constants.}$$

We will classify second order linear DEs as indicated by the following diagram.

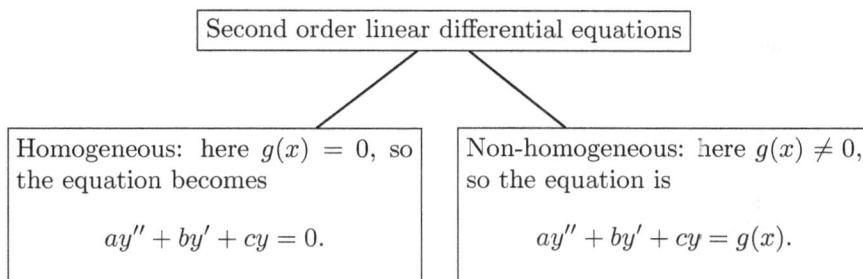

Second order linear differential equations

Homogeneous: here $g(x) = 0$, so the equation becomes

$$ay'' + by' + cy = 0.$$

Non-homogeneous: here $g(x) \neq 0$, so the equation is

$$ay'' + by' + cy = g(x).$$

We say that the DE

$$ay'' + by' + cy = 0, \tag{3.2}$$

is the **homogeneous** DE **associated** to the **non-homogeneous** DE

$$ay'' + by' + cy = g(x) \tag{3.3}$$

For example,

$$y'' - 3y' + 2y = 3x$$

has the associated homogeneous differential equation

$$y'' - 3y' + 2y = 0.$$

When solving a non-homogeneous DE, we actually solve two differential equations separately:

- First, we find solutions y_1 and y_2 to the associated homogeneous DE. These solutions that we need to find are called the **fundamental solution** to the associated homogeneous DE.

- We find a solution to the non-homogeneous DE, that is, a solution to $ay'' + by' + cy = g(x)$ whenever $g(x) \neq 0$. This solution is called a **particular solution** to the DE.

The complete solution to our second order non-homogeneous DE will then be

homogeneous solution + particular solution.

3.1   Linear Homogeneous DE

The first question is, then, how do we find the fundamental solutions y_1 and y_2 of a homogeneous DE. Let's look at an example.

Example 3.1

Find two solutions to the DE

$$y'' - y = 0.$$

Solution: If we write the equation as

$$y'' = y,$$

then the function $y_1(x) = e^x$, whose derivative, of any order, is always itself, may be easily seen to be a solution. With a little bit of more intuition we can guess that $y_2(x) = e^{-x}$ is also a solution to the DE. Indeed, if we realized that for any constants c_1 and c_2,

$$y_1(x) = c_1 e^x \qquad \text{and} \qquad y_2(x) = c_2 e^{-x}$$

are solutions to our DE, then we have found the fundamental solutions in their more general form. So we guessed!

What about guessing the solutions to the DE $y'' + 2y' + y = 0$? In this case, guessing becomes extremely difficult, and not even worth trying. Therefore, a standardized technique is the need of the day.

Let's revisit the equation of example 3.1,

$$y'' - y = 0, \tag{3.4}$$

which has the solutions

$$y = e^x \qquad \text{and} \qquad y = e^{-x}.$$

This means that $y = e^m x$, with $m = 1, -1$, will yield the solutions to the differential equation **(3.4)**. Now let's consider an algebraic equation related to this DE:

$$\underset{\text{Order 2}}{\swarrow} \; y'' \; - \; y \underset{\text{Order 0}}{\searrow} = 0$$

$$\downarrow$$

$$\underset{\text{Degree 2}}{\swarrow} \; m^2 \; - \; 1 \underset{\text{Degree 0}}{\searrow} = 0.$$

The related algebraic equation is a quadratic equation with solutions $m = 1, -1$. Do you see the connection?

So the intuitive argument is:

> A second order linear homogeneous differential equation with constant coefficients has an algebraic equation of second degree associated to it.

Now let's look at the algorithm to solve a second order linear homogeneous DE with constant coefficients:

Algorithm

(1) Let $y = e^{mx}$ be a "trial solution" to the DE

$$ay'' + by' + cy = 0.$$

(2) The DE will turn into the quadratic equation

$$am^2 + bm + c = 0.$$

---

(3) The above is a quadratic equation in the variable m, so we solve the equation for this variable.

(4) If the solutions to the are m_1 and m_2, then the solutions to the DE are $y_1 = e^{m_1 x}$ and $y_2 = e^{m_2 x}$.

(5) Finally, the complete solution to the DE is written as a linear combination
$$y = c_1 y_1 + c_2 y_2 = c_1 e^{m_1 x} + c_2 e^{m_2 x}.$$

Now the next question is: can we prove that the problem of solving $ay'' + by' + cy = 0$ turns into the problem of solving the quadratic equation $am^2 + bm + c = 0$? Let's now look at such proof.

Assume $y = e^{mx}$ is a trial solution. Substitute this trial solution in $ay'' + by' + cy = 0$, which will yield

$$am^2 e^{mx} + bm e^{mx} + c e^{mx} = 0.$$

But all the terms in the above equation have the common factor e^{mx}, and since this is never zero, we can divide by it to obtain

$$am^2 + bm + c = 0,$$

which is the quadratic equation we were looking for. So, the big picture is illustrated by the following diagram

Second order homogeneous DE $\longleftrightarrow$ Quadratic equation

Solutions to the DE $\longleftrightarrow$ Roots of the quadratic equation.

To study the nature of the solutions to the second order homogeneous DE, we need to discuss the nature of the solutions to the associated quadratic equation. Recall that for this we consider the so-called **discriminant**

$$\Delta = b^2 - 4ac$$

of the quadratic equation $am^2 + bm + c = 0$. We have the following possibilities for the roots of this quadratic equation in terms of its discriminant:

Case 1 If $\Delta > 0$, the roots of the quadratic equation are real and distinct.

Case 2 If $\Delta < 0$, the roots of the quadratic equation are complex.

Case 3 If $\Delta = 0$, the roots are real and equal, that is, the quadratic equation has a single root of multiplicity 2.

The DE $y'' - y = 0$ had an associated quadratic equation that falls into case 1. To find solutions to a DE for which the associated quadratic equation falls into case 2, we need to discuss a result called Euler's identity. First let's look at the following example to see what the solutions to a DE look like when we have the case 2.

Example 3.2

Solve the differential equation

$$y'' + y = 0.$$

Solution: Let $y = e^{mx}$ be a trial solution. The associated quadratic equation is

$$m^2 + 1 = 0.$$

Solving for m, we get

$$m_1 = i \qquad \text{and} \, m_2 = -i.$$

Hence, the solutions are imaginary. The solutions to the DE are then

$$y_1 = e^{ix} \qquad \text{and} \qquad y_2 = e^{-ix},$$

and the complete solution is $y = c_1 e^{ix} + c_2 e^{-ix}$. However, we can do a little bit more to this solution. For this, we will need Euler's identity, which we will now discuss.

Euler's Identity   For all $x \in \mathbb{R}$, we have

$$e^{ix} = \cos x + i \sin x.$$

Note then that

$$e^{-ix} = e^{i(-x)} = \cos(-x) + i \sin(-x) = \cos x - i \sin x.$$

We can have *Mathematica* compute this as well:

In[1]:= $\left\{ \texttt{ComplexExpand}\left[e^{t\,i}\right], \texttt{ComplexExpand}\left[e^{-t\,i}\right] \right\}$

Out[1]= $\{\texttt{Cos[t]} + i\,\texttt{Sin[t]}, \texttt{Cos[t]} - i\,\texttt{Sin[t]}\}$

We will apply this formula to the solution of the differential equation in example 3.2,

$$y = c_1 e^{ix} + c_2 e^{-ix}. \tag{3.5}$$

By Euler's identity, we can write equation (**3.5**) as

$$\begin{aligned} y &= c_1(\cos x + i \sin x) + c_2(\cos x - i \sin x) \\ &= (c_1 + c_2) \cos x + i(c_1 - c_2) \sin x, \end{aligned}$$

and if we define $\lambda = c_1 + c_2$ and $\mu = i(c_1 - c_2)$, then the solution (**3.5**) can be written as

$$y = \lambda \cos x + \mu \sin x. \tag{3.6}$$

However, if we consider the general second order homogeneous differential equation with constant coefficients,

$$ay'' + by' + cy = 0,$$

which may give rise to complex roots with both real and imaginary part being nonzero, then the computations are slightly more elaborate than for the one above. Let's look at an example.

Example 3.3

Solve the DE

$$y'' - 4y' + 5y = 0.$$

Solution: The associated differential equation is

$$m^2 - 4m + 5 = 0.$$

We solve this equation for m:

$$m = \frac{4 \pm \sqrt{16 - 4(1)(5)}}{2(1)} = \frac{4 \pm 2i}{2} = 2 \pm i.$$

Unlike in example 3.2, this time the solutions to the quadratic equation have both real and imaginary parts. So we have to study the general case $m_1 = a + bi$ and $m_2 = a - bi$.

Let's consider then the general case for complex roots. If $m_1 = a + bi$ and $m_2 = a - bi$, then the solution to the differential equation is

$$
\begin{aligned}
y &= c_1 y_1 + c_2 y_2 \\
&= c_1 e^{m_1 x} + c_2 e^{m_2 x} \\
&= c_1 e^{(a+bi)x} + c_2 e^{(a-bi)x} \\
&= c_1 e^{ax} e^{bix} + c_2 e^{ax} e^{-bix} \\
&= e^{ax} \left(c_1 e^{bix} + c_2 e^{-bix} \right),
\end{aligned}
$$

and using Euler's identity, we have

$$
\begin{aligned}
e^{bix} &= \cos(bx) + i \sin(bx) \\
e^{-bix} &= \cos(bx) - i \sin(bx).
\end{aligned}
$$

Therefore,

$$
\begin{aligned}
y &= e^{ax} \left[c_1 (\cos(bx) + i \sin(bx)) + c_2 (\cos(bx) - i \sin(bx)) \right] \\
&= e^{ax} \left[(c_1 + c_2) \cos(bx) + i(c_1 - c_2) \sin(bx) \right].
\end{aligned}
$$

If we let $C_1 = c_1 + c_2$ and $C_2 = i(c_1 - c_2)$, then we have

$$y = e^{ax} \left(C_1 \cos(bx) + C_2 \sin(bx) \right). \tag{3.7}$$

Let's use this for the solution of example 3.3:

Example 3.4

Solve the DE

$$y'' - 4y' + 5y = 0.$$

Solution: We have seen that the associated quadratic equation has roots $m_1 = 2 + i$ and $m_2 = 2 - i$. Then, in the context of this problem, $a = 2$ and $b = 1$, so the expression **(3.7)** for the solution yields

$$y = e^{2x} \left(C_1 \cos x + C_2 \sin x \right).$$

Here is a quick recapitulation on the process used to solve a DE that falls into the case 2, like that of example 3.3.

(1) Write down the associated quadratic equation:

$$am^2 + bm + c = 0.$$

(2) Solve the quadratic equation for m to obtain the roots $m = m_1, m_2$.

(3) The solutions are of the form $m_1 = a + bi$ and $m_2 = a - bi$.

(4) Then the solution to the DE is

$$y = e^{ax} \left[C_1 \cos(bx) + C_2 \sin(bx) \right].$$

Now let's consider case 3, that is, the case when the roots of the associated quadratic equation are real and equal. We'll begin by looking at an example.

Example 3.5

Solve the differential equation

$$y'' = 0.$$

Solution: We integrate the DE twice to obtain

$$y = c_2 + c_1 x. \tag{3.8}$$

But we know that any second order differential equation should have two solutions. The function **(3.8)** is one of them, so what's the other? Let's go back to the standard approach of trial solutions, i. e.

$$y = e^m x.$$

Irrespective of the value of m, $y = e^m x$ will yield a solution. This answers the question. Let's compute

$$y' = me^{mx} \qquad \text{and} \qquad y'' = m^2 e^{mx}.$$

Then we substitute this into the DE:

$$y'' = 0 \qquad \Rightarrow \qquad m^2 e^{mx} = 0 \qquad \Rightarrow m^2 = 0.$$

The quadratic equation $m^2 = 0$ has two identical roots $m_1 = 0$ and $m_2 = 0$. Therefore, when we substitute into $y = e^{mx}$ we get

$$y = e^{0x} = e^0 = 1.$$

The point here is that in these cases, the solutions are of the form

$$y_1 = c_2 + c_1 x \qquad \text{and} \qquad e^{mx},$$

and the complete solution is

$$y = y_1 \cdot y_2 = (c_2 + c_1 x)e^{mx}.$$

(The justification that $y = y_1 \cdot y_2$ is a solution is an involved analysis and it is left as a challenging exercise.)

Now let's see the algorithm for this type of second order DE:

> ## Algorithm
>
> (1) Write down the associated quadratic equation:
>
> $$am^2 + bm + c = 0.$$
>
> (2) Solve the quadratic equation for m to obtain the roots $m = m_1, m_2$.
>
> (3) In this case, we will have $m_1 = m_2$ (equal roots).
>
> (4) The solution will then be $y = (c_2 + c_1 x)e^{mx}$.

Now we can combine all the algorithms we have seen into a single one that covers all the three cases regarding the discriminant of the associated quadratic equation.

> ## Algorithm
>
> (1) Find the associated quadratic equation.
>
> (2) Solve the quadratic equation, and look at the values of m_1 and m_2.

(3) If $m_1 \neq m_2$, and both roots are real, then go to step 4.

(4) If $m_1 = m_2$, then go to step 5.

(5) If m_1 and m_2 are complex numbers of the form $a \pm bi$, then go to step 6.

(6) The solution is $y = c_1 e^{m_1 x} + c_2 e^{m_2 x}$.

(7) The solution is $y = (c_2 + c_1 x) e^{mx}$.

(8) The solution is $y = e^{ax}(c_1 \cos(bx) + c_1 \sin(bx))$.

3.2   Linear Independence and the Wronskian

Now we will explore some theoretical ideas that will help us to understand better the results discussed in the previous section.

In the context of second order linear homogeneous DE with constant coefficients,

$$ay'' + by' + cy = 0,$$

it has to be noted that

(1) Two initial conditions are needed.

(2) Two solutions y_1 and y_2 that are obtained must be linearly independent of each other (see section 1.2.9).

(3) y_1 and y_2 need to satisfy the initial conditions.

If all three above are true then $\{y_1, y_2\}$ is called a fundamental set of solutions. Moreover, the solution is represented in the form

$$y = c_1 y_1 + c_2 y_2.$$

Why? The answer is the following result, called the **principle of superposition**.

Theorem 3.6 (Principle of superposition)
If y_1 and y_2 are nontrivial solutions to

$$ay'' + by' + cy = 0,$$

then $y = c_1 y_1 + c_2 y_2$ is also a solution to this DE.

Proof: Given that y_1 and y_2 are solutions to the DE, we have

$$ay_1'' + by_1' + cy_1 = 0 \qquad \text{and} \qquad ay_2'' + by_2' + cy_2 = 0.$$

Now, our goal is to show that $y = c_1 y_1 + c_2 y_2$ is also a solution. We are going to use this value of y in the left-hand side of the DE and show that this is equal to zero:

$$
\begin{aligned}
a(c_1 y_1 + c_2 y_2)'' + b(c_1 y_1 + c_2 y_2)' + c(c_1 y_1 + c_2 y_2) &= c_1(a y_1'' + b y_1' + c y_1) \\
&+ c_2(a y_2'' + b y_2' + c y_2) \\
&= c_1(0) + c_2(0) \\
&= 0.
\end{aligned}
$$

Hence, $y = c_1 y_1 + c_2 y_2$ is also a solution to the DE.   ∎

3.2.1   Linear Independence

In section 1.2.9, we have discussed the concepts of linear dependence and linear independence. Such concepts were discussed in a rather general context, so here we will revisit these ideas in the context of second order differential equations.

If y_1 and y_2 are two functions, then definition 1.27 tells us that y_1 and y_2 are linearly dependent if there are constants c_1 and c_2, not both zero, such that

$$c_1 y_1 + c_2 y_2 = 0.$$

Here is another way to look at linear dependence. Suppose that the relation above is true for some constants c_1 and c_2. If at least one of these constants is not zero, we may assume, without loss of generality, that it is $c_1 \neq 0$. Look, then that we can solve the relation for y_1 as follows:

$$y_1 = -\frac{c_2}{c_1} y_2,$$

so y_2 is a constant multiple of y_1, that is, y_2 can be obtained from y_1 by multiplying y_1 by the constant $-\frac{c_2}{c_1}$.

This means that finding two linearly dependent solutions y_1 and y_2 to a DE is rather a redundant result, since one solution is simply the multiple of the other. What we are interested in is solutions that are not related to each other so much, that is, in solutions that are linearly independent. When we find such solutions, then we can say that we have found a fundamental set of solutions to the DE.

3.2.2   The Wronskian

The Wronskian is just a determinant that will allow us to figure out whether two functions are linearly independent or not. Let's see what this determinant is.

Definition 3.7

Let y_1 and y_2 be two functions of x. The Wronskian W of y_1 and y_2 is also a function of x, and it is given by

$$W(y_1, y_2)(x) = \begin{vmatrix} y_1(x) & y_2(x) \\ y_1'(x) & y_2'(x) \end{vmatrix} = y_1(x)y_2'(x) - y_2(x)y_1'(x).$$

The following theorem will tell us how to use the Wronskian to test two functions for linear independence.

Theorem 3.8

Let y_1 and y_2 be two functions of x. If

$$W(y_1, y_2)(x) \neq 0$$

*for some $x = x_0$, then y_1 and y_2 are linearly independent. If y_1 and y_2 are solutions to a second order differential equation, then $\{y_1, y_2\}$ is called a **fundamental set of solutions** to the DE.*

This approach of using the Wronskian of two functions is a much faster way to test for linear independence than the idea based on the definition of linear independence (see example 1.31). Let's see an example.

Example 3.9

Let $y_1 = \cos x$ and $y_2 = \sin x$ be two solutions to a second order DE. Show that $\{y_1, y_2\}$ is a fundamental set of solutions to the DE.

Solution: We will use the Wronskian. We compute $y_1' = -\sin x$ and $y_2' = \cos x$, so the Wronskian of y_1 and y_2 is

$$W(y_1, y_2)(x) = \begin{vmatrix} \cos x & \sin x \\ -\sin x & \cos x \end{vmatrix} = \cos^2 x + \sin^2 x = 1 \neq 0.$$

Hence, the functions y_1 and y_2 are linearly independent, and $\{y_1, y_2\}$ is a fundamental set of solutions to the DE.

We can have *Mathematica* compute the Wronskian. For example, to compute the one in example 3.9, we evaluate

```
In[1]:= Wronskian[{Cos[x], Sin[x]}, x]

Out[1]= 1
```

Now we will prove the link between the Wronskian and the idea of linear independence, that is, we will prove theorem 3.8.

Proof (theorem 3.8): . Let's begin with the expression

$$c_1 y_1 + c_2 y_2 = 0. \tag{3.9}$$

Just for fun, let's differentiate both sides of the above equation to obtain

$$c_1 y_1' + c_2 y_2' = 0. \tag{3.10}$$

We can rewrite equation **(3.9)** as

$$c_1 = -c_2 \frac{y_2}{y_1}, \tag{3.11}$$

and we can also rewrite equation **(3.10)** as

$$c_1 = -c_2 \frac{y_2'}{y_1'}. \tag{3.12}$$

From **(3.11)** and **(3.12)**, we have

$$
\begin{aligned}
c_2 \frac{y_2}{y_1} &= c_2 \frac{y_2'}{y_1'} \\
\Rightarrow \qquad c_2 y_2 y_1' &= c_2 y_2' y_1 \\
\Rightarrow \quad c_2 y_2' y_1 - c_2 y_2 y_1' &= 0 \\
\Rightarrow \quad c_2 (y_2' y_1 - y_2 y_1') &= 0.
\end{aligned}
$$

Note that the expression multiplying c_2 in the left-hand side of the last equation is the Wronskian of y_1 and y_2. Therefore, we have

$$c_2 W(y_1, y_2) = 0.$$

For the above equation to hold, we need that either c_2 or $W(y_1, y_2)$ be zero. If we find that $W(y_1, y_2)(x)$ is not zero at some x, then this will force c_2 to be the one that is zero, that is,

$$W(y_1, y_2) \neq 0 \qquad \text{implies} \qquad c_2 = 0.$$

By a similar process we can show that also

$$W(y_1, y_2) \neq 0 \qquad \text{implies} \qquad c_1 = 0.$$

This proves what we wanted, for if the Wronskian is not zero at some point, this implies that c_1 and c_2 in equation **(3.9)** are both zero, which tells us that y_1 and y_2 are linearly independent. ∎

3.3   Solving $ay'' + by' + cy = g(x)$

Recall that

$$ay'' + by' + cy = g(x)$$

is a second order differential equation, and when we seek a solution to this DE, we do it in two parts.

(1) First we deal with $ay'' + by' + cy = 0$, and the solution is of the form

$$y_h = c_1 y_1(x) + c_2 y_2(x).$$

(2) We find a particular solution, y_p, that solves the DE

$$ay'' + by' + cy = g(x).$$

There are two commonly used approaches to finding y_p: the method of undetermined coefficients and the method of variation of parameters.

$$\text{Methods to find } y_p \begin{cases} \text{Method of undetermined coefficients.} \\ \text{Method of variation of parameters.} \end{cases}$$

3.3.1 The Method of Undetermined Coefficients

Let's begin with an example. Consider the differential equation

$$y'' - 5y' + 4y = e^{7x}.$$

Let's solve the associated homogeneous equation,

$$y'' - 5y' + 4y = 0.$$

The solution is $y_h = c_1 e^{4x} + c_2 e^x$. Now we will find the particular solution y_p to the non-homogeneous DE. Assume y_p to be a trial solution by setting

$$y_p = A e^{7x},$$

where A is a coefficient not known to us.

Remark
To find y_p it is imperative to find the value of the undetermined coefficient A. Hence the name of the method.

Since y_p is a trial solution, it must satisfy

$$y'' - 5y' + 4y = e^{7x}.$$

We have that

$$y_p' = 7Ae7x \qquad y_p'' = 49Ae^{7x}.$$

Therefore, we must have

$$
\begin{aligned}
49Ae^{7x} - 5 \cdot 7Ae^{7x} + 4Ae^{7x} &= e^{7x} \\
\Rightarrow \qquad (49A - 35A + 4A)e^{7x} &= e^{7x} \\
\Rightarrow \qquad 18Ae^{7x} &= e^{7x} \\
\Rightarrow \qquad 18A &= 1 \\
\Rightarrow \qquad A &= \frac{1}{18}.
\end{aligned}
$$

Hence, the particular solution is

$$y_p = \frac{1}{18}e^{7x},$$

and the final solution to the non-homogeneous DE is

$$y = y_h + y_p = c_1 e^{4x} + c_2 e^{x} + \frac{1}{18}e^{7x}.$$

Before we explain more on the details of this method, let's solve again the above differential equation using *Mathematica*.

Example 3.10

Follow the procedure given above to find a particular solution y_p to the differential equation

$$y'' - 5y' + 4y = e^{7x}$$

using *Mathematica*

Solution: First, let's give a name to this DE in *Mathematica*:

In[1]:= **Clear[eq, y, x]; eq = (y')'[x] - 5 y'[x] + 4 y[x] == e$^{7\,x}$**

Out[1]:= 4 y[x] - 5 y'[x] + y''[x] == e$^{7\,x}$

As above, we figure out a reasonable trial solution looking at the right-hand side of the DE. The trial solution will be $y_p = Ae^{7x}$, so we set

In[2]:= **Clear[yp]; yp[x_] := A e$^{7\,x}$**

Out[2]:=

We only need to substitute **y[x]** by **yp[x]** into the DE by a replacement in *Mathematica*:

In[3]:= **eq /. {y → yp}**

Out[3]:= 18 A e$^{7\,x}$ == e$^{7\,x}$

From this, you can easily get the value of the undetermined coefficient A, but let's ask *Mathematica* to do this too.

In[4]:= **Clear[ucff]; ucff = Solve[eq /. {y → yp}, A]**

Out[4]:= $\left\{\left\{A \to \frac{1}{18}\right\}\right\}$

Hence, the particular solution is

In[5]:= **yp[x] /. ucff**

Out[5]:= $\left\{\frac{e^{7\,x}}{18}\right\}$

as before.

Now some quick facts on solving the DE $ay'' + by' + cy = g(x)$.

Quick Facts

(1) Find the solution y_h to the associated homogeneous DE

$$ay'' + by' + cy = 0.$$

(2) Seek a trial solution y_p. Choose y_p based on the nature of $g(x)$ on the right-hand side of the DE.

(3) y_p will contain $A \times$ (some function), where A is the undetermined coefficient. Sometimes we may have more than one undetermined coefficient.

(4) Solve for the undetermined coefficient A.

(5) The final solution to the DE will be $y = y_h + y_p$.

Now the question is, how do we choose the trial solution to find y_p? Let's look at some possible cases:

How to choose y_p

(1) If $g(x)$ is of the form e^{mx}, where m is a constant, then choose $y_p = Ae^{mx}$.

(2) If $g(x)$ is a polynomial in x, that is,

$$g(x) = a_n x^n + a_{n-1} x^{n-1} + \cdots + a_1 x + a_0.$$

then choose y_p to be

$$y_p = A_n x_n + A_{n-1} x_{n-1} + \cdots + A_1 x + A_0.$$

For example, if $g(x) = 3x^2 - 2x + 1$, then choose $y_p = Ax^2 + Bx + C$.

(3) If $g(x)$ is $\sin x$ or $\cos x$, then choose $y_p = A \sin x + B \cos x$.

Let's look at an example.

Example 3.11

Solve the differential equation

$$y'' + 4y' - 12y = \sin(2x).$$

Solution: The solution to the associated homogeneous DE is

$$y_h = c_1 e^{-6x} + c_1 e^{2x}.$$

To find y_p, what is the guess? The trial solution we should choose is

$$y_p = A \sin(2x) + B \cos(2x).$$

Then we have

$$y'_p = 2A\cos(2x) - 2A\sin(2x) \qquad \text{and} \qquad y''_p = -4A\sin(2x) - 4B\cos(2x).$$

Now we substitute these values into the DE:

$$-4A\sin(2x) - 4B\cos(2x) \quad + \quad 4(2A\cos(2x) - 2B\sin(2x))$$
$$- \quad 12(A\sin(2x) + B\cos(2x) = \sin(2x).$$

Simplifying the RHS of this equation, we get

$$[-4A + 8B - 12A]\sin(2x) + [-4B - 8A - 12B]\cos(2x) = \sin(2x) + 0\cos(2x).$$

The next step is crucial. In order to have the above equation true, the coefficients of $\sin(2x)$ and $\cos(2x)$ in the LHS and the RHS must be the same:

$$-16A + 8B = 1$$
$$-16B - 8A = 0.$$

This implies that $A = -2B$ and $B = \frac{1}{40}$, so $A = -\frac{1}{20}$. Hence, the particular solution is

$$y_p = -\frac{1}{20}\sin(2x) + \frac{1}{40}\cos(2x),$$

and the final solution is

$$y = y_h + y_p = c_1 e^{-6x} + c_1 e^{2x} - \frac{1}{20}\sin(2x) + \frac{1}{40}\cos(2x).$$

Exercise: If instead of $y_p = A\sin(2x) + B\cos(2x)$ we had guessed $y_p = A\sin(2x)$, what would have happened?

Example 3.12

Using *Mathematica*, find a particular solution to the DE of example 3.11.

Solution: We set

In[1]:= **Clear[x, y, eq]; eq = (y')'[x] + 4 y'[x] - 12 y[x] == Sin[2 x]**

Out[1]:= **-12 y[x] + 4 y'[x] + y''[x] == Sin[2 x]**

Now we choose a trial particular solution with undetermined coefficients. We know that such trial solution is $y_p(x) = A\sin(2x) + B\cos(2x)$, so we define

In[2]:= **Clear[yp]; yp[x_] := A Sin[2 x] + B Cos[2 x]**

Out[2]:=

We substitute this into the DE and assign the result to **seq**:

In[3]:= **seq = eq /. {y → yp}**

Out[3]:= **-4 B Cos[2 x] - 4 A Sin[2 x] - 12 (B Cos[2 x] + A Sin[2 x]) +**
    4 (2 A Cos[2 x] - 2 B Sin[2 x]) == Sin[2 x]

Here is a nice way to ask *Mathematica* to find the coefficients of $\sin(2x)$ and $\cos(2x)$ in both sides of the equation and match them.

In[4]:= **system = Thread[(Coefficient[#1, {Cos[2 x], Sin[2 x]}] &) /@ seq]**

Out[4]= {8 A - 16 B == 0, -16 A - 8 B == 1}

Now we solve the system,

In[5]:= **cffs = Solve[system, {A, B}]**

Out[5]= $\left\{\left\{A \rightarrow -\frac{1}{20}, B \rightarrow -\frac{1}{40}\right\}\right\}$

Hence, the particular solution is

In[6]:= **yp[x] /. cffs**

Out[6]= $\left\{-\frac{1}{40} \text{Cos}[2\,x] - \frac{1}{20} \text{Sin}[2\,x]\right\}$

Let's do a quick recapitulation on the method of undetermined coefficients, and then discusses some possible issues that may arise when using this method to solve a DE.

For $ay'' + by' + cy = g(x)$, the solution is

$$y = y_h + y_p,$$

where y_h is the solution to the associated homogeneous DE, and y_p is a particular solution to the non-homogeneous DE, which is found by the method of undetermined coefficients.

Remark

It is always a good practice to find the homogeneous solution to the DE even if we were not asked for it. The reason is that

the homogeneous solution sometimes affects the choice of the particular solution y_p. If in our initial guess, y_p contains a part of the homogeneous solution, we need to modify our guess.

Let's look at an example that will illustrate the situation described in the above remark.

Example 3.13

Solve the DE

$$y'' - 4y' - 12y = e^{-2x}.$$

Solution: The homogeneous solution is $y_h = c_1 e^{-2x} + c_2 e^{6x}$. Now we seek for the particular solution y_p. Our initial guess would be

$$y_p = Ae^{-2x}.$$

But note that this choice of y_p is contained in the homogeneous solution y_h, so this may not work. So, we change our guess to

$$y_p = Axe^{-2x}.$$

Then we have

$$y_p' = A\left(e^{-2x} - 2xe^{-2x}\right) \qquad y_p'' = A\left(4xe^{-2x} - 4e^{-2x}\right).$$

Substituting into the DE, we have

$$A\left(4xe^{-2x} - 4e^{-2x}\right) - 4A\left(e^{-2x} - 2xe^{-2x}\right) - 12xe^{-2x} = e^{-2x}.$$

Simplifying the RHS, we get

$$(-8A)e^{-2x} = e^{-2x},$$

so we must have $A = \frac{-1}{8}$. Therefore, the particular solution is

$$y_p = -\frac{1}{8}xe^{-2x},$$

and the final solution to the DE is

$$y = y_h + y_p = c_1 e^{-2x} + c_2 e^{6x} - \frac{1}{8}xe^{-2x}.$$

Now, we would like you to note the following:

(1) To find y_p, you should guess carefully, noting the form of y_h and acting accordingly.

(2) The process of finding y_p is more algebra than anything else.

And finally, realize the limitation of the method of undetermined coefficients:

Limitation The method of undetermined coefficients is applicable to a very limited class of functions in the RHS, that is, the cases when $g(x)$ is an exponential function, a trigonometric function, or a polynomial function.

When we have a combination of polynomial, exponential or trigonometric functions, our guess is also a combination of the corresponding guesses for each of the functions involved. Let's see an example.

Example 3.14

What is the guess for the particular solution to the DE

$$y'' - 4y' - 12y = xe^x.$$

Solution: The homogeneous solution is $y_h = c_1 e^{-2x} + c_2 e^{6x}$. Our guess for the particular solution is

$y_p <$ for x, we need a standard polynomial of degree 1: $Ax + B$

for e^x, we need the guess Ce^x.

So, the actual guess is

$$y_p = (Ax + B)Ce^x.$$

3.3.2 Variation of Parameters

For a second order non-homogeneous DE

$$a(x)y'' + b(x)y' + c(x)y = g(x),$$

this is a more general method to find a particular solution y_p. This method gives us that

$$y_p = -y_1(x) \int \frac{y_2(x)g(x)}{W(y_1, y_2)(x)} \, dx + y_2(x) \int \frac{y_1(x)g(x)}{W(y_1, y_2)(x)} \, dx,$$

provided that $\{y_1, y_2\}$ is a fundamental set of solutions to the associated homogeneous DE

$$a(x)y'' + b(x)y' + c(x)y = 0,$$

that is, $W(y_1, y_2) \neq 0$.

Pros and Cons

(1) Compared to the method of undetermined coefficients, this method applies to a variety of functions for $g(x)$.

(2) We must have $W(y_1, y_2) \neq 0$, otherwise the method doesn't work.

(3) The integrals involved sometimes are too complicated.

(4) It's mandatory to find the homogeneous solution y_h first.

The next question is, why is this method called variation of parameters?

Since it is absolutely necessary to find y_h first, let's look at the solution to

$$ay'' + by' + cy = 0,$$

which is of the form $y_h = c_1 y_1 + c_2 y_2$. To get at the bottom of the problem, c_1 and c_2 are no longer treated as constants when we seek a particular solution y_p. More over, it can be shown that in that case, we must have

$$c_1 = -y_1(x) \int \frac{y_2(x)g(x)}{W(y_1, y_2)(x)} \, dx$$

and

$$c_2 = y_2(x) \int \frac{y_1(x)g(x)}{W(y_1, y_2)(x)} \, dx.$$

Let's look at an example.

Example 3.15

Solve the DE

$$y'' + 9y = \tan(3t).$$

Solution: First, we find the homogeneous solution, which is

$$y_h = c_1 \cos(3t) + c_2 \sin(3t),$$

and from this we obtain $y_1 = \cos(3t)$ and $y_2 = \sin(3t)$. Now we check that $W(y_1, y_2) \neq 0$. Indeed,

$$W(y_1, y_2) = \begin{vmatrix} \cos(3t) & \sin(3t) \\ -3\sin(3t) & 3\cos(3t) \end{vmatrix} = 3\cos^2(3t) + 3\sin^2(3t) = 3 \neq 0.$$

We can then proceed to evaluate

$$c_1 = -\int \frac{y_2(t)g(t)}{W(y_1, y_2)(t)} \, dt = -\int \frac{\sin(3t)\tan(3t)}{3} \, dt.$$

and

$$c_2 = \int \frac{y_1(t)g(t)}{W(y_1, y_2)(t)} \, dt = \int \frac{\cos(3t)\tan(3t)}{3} \, dt.$$

If we evaluate these integrals, we will find that

$$c_1 = -\frac{1}{9} \ln|\tan(3t) + \sec(3t)| - \frac{1}{9}\sin(3t)$$

and

$$c_2 = -\frac{1}{9}\cos(3t),$$

so the particular solution is

$$y_p = \left[-\frac{1}{9}\ln|\tan(3t) + \sec(3t)| - \frac{1}{9}\sin(3t)\right]\sin(3t) - \frac{1}{9}\cos^2(3t).$$

The final answer, which you can give explicitly, is found by substituting what we have obtained into

$$y = y_h + y_p.$$

Let's look at another example.

Example 3.16

Solve the DE

$$y''(x) - 4y'(x) + 4y(x) = \frac{e^{2x}}{\sqrt{x}}.$$

Solution: The associated homogeneous DE is

$$y''(x) - 4y'(x) + 4y(x) = 0.$$

We leave it to you to solve this homogeneous differential equation. You will find that the solution is

$$y_h(x) = C_1 e^{2x} + C_2 x e^{2x}.$$

(This is a case where the auxiliary quadratic equation for the homogeneous DE has a repeated root). Next step is to find the Wronskian of the two linearly independent solutions $y_1 = e^{2x}$ and $y_2 = xe^{2x}$ that form the homogeneous solution y_h. We have that

$$W(y_1, y_2) = \begin{vmatrix} y_1 & y_2 \\ y_1' & y_2' \end{vmatrix} = \begin{vmatrix} cce^{2x} & xe^{2x} \\ 2e^2x & 2xe^{2x} + e^{2x} \end{vmatrix} = e^{2x}\left(2xe^{2x} + e^{2x}\right) - 2xe^{2x}e^{2x} = e^{4x}.$$

The method of variation of parameters tells us that the particular solution to our original DE is

$$\begin{aligned}y_p(x) &= -y_1 \int \frac{y_2 g(x)}{W(y_1, y_2)}\,dx + y_2 \int \frac{y_1 g(x)}{W(y_1, y_2)} \\ &= -e^{2x}\int \frac{xe^{2x}}{e^{4x}}\frac{e^{2x}}{\sqrt{x}}\,dx + xe^{2x}\int \frac{e^{2x}}{e^{4x}}\frac{e^{2x}}{\sqrt{x}}\,dx \\ &= -e^{2x}\left(-\frac{2}{3}e^{2x}x^{3/2}\right) + xe^{2x}\left(2x^{1/2}\right) \\ &= 4/3e^{2x}x^{3/2}.\end{aligned}$$

Hence, the final solution to the DE is

$$y = y_h + y_p = c_1 e^{2x} + c_2 x e^{2x} + \frac{4}{3}e^{2x}x^{3/2}.$$

Example 3.17

Solve the DE of example 3.16 using *Mathematica*.

Solution: Let's define

$In[1]:=$ `Clear[deq, x, y]; deq = (y')'[x] - 4 y'[x] + 4 y[x] ==` $\frac{e^{2x}}{\sqrt{x}}$

$Out[1]:=$ $4 y[x] - 4 y'[x] + y''[x] == \frac{e^{2x}}{\sqrt{x}}$

In this example, we will allow ourselves to use the sophisticated *Mathematica* function **DSolve** to find the solution to the associated homogeneous equation immediately:

$In[2]:=$ `sols = List @@ (y[x] /. DSolve[deq[[1]] == 0, y[x], x])[[1]];`
`y = sols /. {C[_] → 1}`

$Out[2]:=$ $\left\{ e^{2x}, \ e^{2x} x \right\}$

Let's give a name to our forcing function (the right-hand side of the DE):

$In[3]:=$ `g[x_] := deq[[2]]`

$Out[3]:=$

We also need the Wronskian of our homogeneous solutions,

$In[4]:=$ `w = Wronskian[y, x]`

$Out[4]:=$ e^{4x}

Since the Wronskian is not zero, we can proceed to compute the integrals that yield the particular solution,

$In[5]:=$ `yp = -y[[1]]` $\int \frac{y[[2]] \, g[x]}{w} \, dx$ `+ y[[2]]` $\int \frac{y[[1]] \, g[x]}{w} \, dx$

$Out[5]:=$ $\frac{4}{3} \, e^{2x} x^{3/2}$

Hence, the final solution is

$In[6]:=$ `Plus @@ Append[sols, yp]`

$Out[6]:=$ $\frac{4}{3} \, e^{2x} x^{3/2} + e^{2x} C[1] + e^{2x} x C[2]$

Exercises

(1) Simple equivalence problems. Convert each of the following ODEs into the equivalent algebraic equation (auxiliary equation), solve it and find the roots, and finally characterize the roots. (Refresh factoring skills if needed).

    (a)   i. $y'' - 7y' + 10y = 0$.

    ii. $y'' - 11y' + 28y = 0$.

    iii. $y'' - 19y' + 90y = 0$.

    iv. $y'' - 27y = 0$.

(b)   i. $y'' - 2y' + y = 0$.

    ii. $y'' - 22y' + 121y = 0$.

    iii. $y'' + 8y' + 16y = 0$.

(c)   i. $2y'' + 32y = 0$.

    ii. $y'' - 2y' + 10y = 0$.

    iii. $y'' + 3y = 0$.

(2) For each of the problems in (1), use the appropriate representation to write the solution.

(3) **Linear independence/dependence of the solutions.** For each problem in (1), verify that the solutions y_1 and y_2 are linearly independent.

(4) **Handling non-homogeneous DE.**

  (a) **The method of undetermined coefficients.** Using the method of undetermined coefficients, solve the following non-homogeneous DEs. (Note: The homogeneous solution and the particular solution together describe the complete solution.)

       i. $y'' - 7y' + 10y = e^{5x}$.

       ii. $y'' - 7y' + 10y = e^{9x}$.

       iii. $y'' - 5y' + 4y = e^{7x}$, $y(0) = 0$, $y'(0) = 0$.

       iv. $y'' - 11y' + 28y = \sin(3x)$, $y(0) = L$, $y'(0) = \dfrac{K}{K^2 + L^2}$, where L and K are positive constants.

       v. $y'' - 19y' + 90y = (x^2 + 1)e^x$.

       vi. $y'' - 27y = 3x^2 - x^2 + 1$.

  (b) **The method of variation of parameters.** Using the method of variation of parameters, solve the following DEs completely.

       i. $y'' + 16y = \tan(4x)$.

       ii. $y'' - 2y' + y = e^{2x} \ln x$. (Can you apply the method of undetermined coefficients here? If not, explain why.)

       iii. $y'' + y = \sec x$.

       iv. $y'' - 2y' + y = e^{-x}$.

(5) **Challenging problems.**

  (a) **Abel's identity.** The Wronskian $W(y_1, y_1)(x)$ of two solutions to the homogeneous ODE

$$a(x)y'' + b(x)y' + c(x)y = 0$$

satisfies
$$a(x)W'(y_1, y_2)(x) + b(x)W(y_1, y_2)(x) = 0,$$
where $W(y_1, y_2)(x) = y_1(x)y_2'(x) - y_1'(x)y_2(x)$. Show that W satisfies the identity
$$W(y_1, y_2)(x) = W(x_0)e^{-\int_{x_0}^x \frac{b(t)}{a(t)} dt}.$$

(b) Let a, b and c be constants, $a \neq 0$, and f be a continuous function in a neighborhood of $x = x_0$. Let y_1 and y_2 be two linearly independent solutions of $ay'' + by' + cy = 0$. Define $W(y_1, y_2)(x) = y_1(x)y_2'(x) - y_1'(x)y_2(x)$. Then the DE
$$ay'' + by' + cy = 0 = f(x)$$
has the particular solution
$$y_p(x) = -\frac{y(x)}{a}\int \frac{y_2(x)f(x)}{W(y_1, y_2)(x)} dx + \frac{y_2(x)}{a}\int \frac{y_1(x)f(x)}{W(y_1, y_2)(x)} dx$$

(c) **Reduction of order.** For a variable coefficient DE,
$$a(x)y'' + b(x)y' + c(x)y = 0. \tag{3.13}$$

Reduction of order is a useful method. The method requires one solution to be known. Please read the method carefully.

  i. Let $y_1(x) = \ldots$ be a known solution.
 ii. Assume $y_2(x) = v(x)y_1(x)$ for an appropriate choice of $v(x)$.
iii. Then
$$\begin{aligned} y_2'(x) &= v'y_1 + vy_1' \\ y_2''(x) &= v''y_1 + v'y_1' + v'y_1' + vy_1'' \\ &= v''y_1 + 2v'y_1' + vy_1''. \end{aligned}$$

 iv. Since y_2 is a solution of **(3.13)**,
$$a(x)y_2'' + b(x)y_2' + c(x)y = 0,$$
plugging in for y_2, y_2' and y_2'', a new DE in terms of v will be obtained, which will contain v' and v'' (no v).
  v. Set $w = v' \implies w' = v''$, so a new equation in terms of w is obtained.
 vi. Solve for w, then get v, and from this finally get y_2.

Use the method to solve
$$x^2y'' + 2xy' - 2t = 0,$$
given that $y_1(x) = x$ is a known solution.

(6) **Mechanical vibrations.** The mechanical vibration of a spring is modeled by the DE

$$my'' + \gamma y' + ky = F(x), \qquad y(0) = y_0, \ y'(0) = y_0',$$

where

- F is an external force,
- $-\gamma y'$ is the damping force,
- ky is the spring force,
- y is the distance measured from the equilibrium point,
- y_0 is the initial displacement from equilibrium,
- y_0' is the initial velocity,
- $mg = KL$, where L is the amount of stretch by attaching an object of mass m at the end of the spring.

(a) In the case of free, undamped vibration ($F = 0$ and $\gamma = 0$), show that the solution to the DE can be represented as

$$y(x) = R\cos(\omega_0 x - \delta),$$

where $R = \sqrt{c_1^2 + c_2^2}$, (c_1, c_2 are constants of integration), $\delta = \arctan\left(\dfrac{c_2}{c_1}\right)$, and $\omega_0 = \sqrt{\dfrac{K}{m}}$.

(b) In the case forced, undamped vibrations ($\gamma = 0$), where $F(x)$ is described as $F(x) = F_0 \cos x$, assuming $\omega \neq \omega_0$ and using the method of undetermined coefficients, show that the solution $y(x)$ (displacement) is given by

$$y(x) = R\cos(\omega_0 x - \delta) + \frac{F_0}{m(\omega_0^2 - \omega_2)}\cos(\omega x).$$

What is wrong if $\omega = \omega_0$?

(7) A standard electric circuit with a capacitor of capacitance C, a resistor of resistance R and inductor of inductance L follows a second order non-homogeneous DE,

$$L\frac{d^2Q}{dt^2} + R\frac{dQ}{dt} + \frac{1}{C}Q = E(t),$$

where Q is the charge in the capacitor, and $I = \dfrac{dQ}{dt}$ is the current. Assume the initial condition to be $Q(0) = Q_0$, $Q'(0) = I_0$, and also that $E(t) = E_0 \sin(\omega t)$, and that none of the constants ω, L, R, C and E_0 are equal to zero. Find an expression for Q, the charge at any time t. What happens when $t \to \infty$. Use *Mathematica* to plot $Q(t)$.

Chapter 4

Systems of Differential Equations

> In order to understand the ideas of this chapter, you have to make sure you understand the ideas and results on matrices and systems of linear equations introduced in the second part of Chapter 1, which you may consider necessary to revisit before starting this chapter.

Beyond first and second order DEs, there are ample examples found in engineering applications that are extremely complicated. What this means is those problems pose a challenge. For example, how do we solve a seventh order differential equation like

$$2y^{(7)} - 3y^{(4)} + 5y^{(3)} - y' = 0.$$

The analytical techniques that we have learned exhaust themselves at this point. So, we propose to look at the following problem: *an n-th order differential equation with initial conditions.*

Background required

(1) Understand systems of algebraic equations.

(2) Understand a matrix system and the various properties of matrices.

(3) Understand and use eigenvalues and eigenvectors.

Advantages The reason to study systems of DEs (or for that matter any system of equations) is that we can move over to a computer algebra system that will take care of the solution process efficiently and quickly.

4.1   Setting up a System of DEs

Let's look at the following DE.

$$2y'' - 5y' + y = 0. \tag{4.1}$$

Step 1 Introduce new variables,

$$x_1(t) = y(t) \qquad \text{and} \qquad x_2(t) = y'(t)$$

Step 2 Relate y, y', y'' to the new variables (we need to change everything in terms of x_1 and x_2.)

$$y''(t) = x_2'(t).$$

Step 3 Also,

$$x_1'(t) = y'(t) = x_{(}t) \qquad \text{and} \qquad y''(t) = x_2'(t).$$

From the original DE, we have

$$y'' = \frac{5}{2}y' - \frac{1}{2}y,$$

so

$$x_2' = \frac{5}{2}x_2(t) - \frac{1}{2}x_1(t).$$

Do we have a system of first order DEs? Yes:

$$
\begin{aligned}
x_1'(t) &= x_2(t) \\
x_2'(t) &= \frac{5}{2}x_2(t) - \frac{1}{2}x_1(t).
\end{aligned}
$$

Step 4 Set this as a matrix system:

$$\begin{pmatrix} x_1 \\ x_2 \end{pmatrix}' = \begin{pmatrix} 0 & 1 \\ -\frac{1}{2} & \frac{5}{2} \end{pmatrix} \begin{pmatrix} x_1 \\ x_2 \end{pmatrix}.$$

The set-up is complete. Now let $\mathbf{x} = \begin{pmatrix} x_1 \\ x_2 \end{pmatrix}$ so that the system can be represented as

$$\mathbf{x}' = \mathbf{A}\mathbf{x}, \qquad \text{where } \mathbf{A} = \begin{pmatrix} 0 & 1 \\ -\frac{1}{2} & \frac{5}{2} \end{pmatrix}.$$

Let's look at another example.

Example 4.1

Set up the system of first order DEs corresponding to the third order DE

$$y^{(3)} - 4y'' + 6y' - y = 0.$$

Solution: (1) Define the variables

$$x_1(t) \ = \ y(t) \tag{4.2}$$
$$x_2(t) \ = \ y'(t) \tag{4.3}$$
$$x_3(t) \ = \ y''(t) \tag{4.4}$$

(2) From **(4.2)** and **(4.3)**, we have

$$x_1' = y' = x_2,$$

and from **(4.3)** and **(4.4)**,

$$x_2' = y'' = x_3.$$

Also, from the original differential equation, we have

$$y^{(3)} = 4y'' - 6y' + y = 4x_3 - 6x_2 + x_1,$$

so the matrix representation of this system is

$$\begin{pmatrix} x_1 \\ x_2 \\ x_3 \end{pmatrix}' = \begin{pmatrix} 0 & 1 & 0 \\ 0 & 0 & 1 \\ 1 & -6 & 4 \end{pmatrix} \begin{pmatrix} x_1 \\ x_2 \\ x_3 \end{pmatrix}$$

4.2   Solving a System of First Order DEs

So far, we have seen how to set up a higher order DE as a system of first order differential equations. Let's set a few norms straight before we study the solution process.

Remark

  (1) We'll look at 2×2 systems only, i.e. the corresponding matrix will be a 2×2 matrix.

  (2) The eigenvalues and eigenvectors of the matrix describe the solutions to the system.

  (3) A 2×2 homogeneous system would be studied in comparison with a second order DE.

  (4) A further classification of fact (3) is like the characterization of roots of a second order DE. Eigenvalues could be classified into the following categories.

$$\text{Categories} \begin{cases} \text{real and distinct} \\ \text{real and equal} \\ \text{complex,} \end{cases}$$

and just like the second order DE, in Chapter 3, the solutions to the system will vary accordingly.

Let's study the solution: Let's start from

$$\mathbf{x}' = \mathbf{A}\mathbf{x}. \tag{4.5}$$

Compare this with a scalar first order DE:

$$x' = ax. \tag{4.6}$$

The point is: knowledge of a solution to **(4.6)** could be used to solve **(4.5)**. How do we solve **(4.6)**?

$$
\begin{aligned}
\int \frac{dx}{x} &= \int a\,dt \\
\ln x &= at + \ln C \\
x &= Ce^{at}.
\end{aligned}
$$

Claim: The solution to **(4.5)** is

$$\mathbf{x} = \mathbf{v}e^{\lambda t}.$$

The column vector $\mathbf{v}$ replaces the constant of integration C, and it's called the eigenvector associated with the eigenvalue λ.

We have $\mathbf{x}' = \lambda \mathbf{v}e^{\lambda t}$. Therefore, from **(4.5)**,

$$\lambda \mathbf{v}e^{\lambda t} = \mathbf{A}\mathbf{v}e^{\lambda t},$$

so

$$0 = \mathbf{A}\mathbf{v}e^{\lambda t} - \lambda \mathbf{v}e^{\lambda t},$$

and factoring, we have

$$0 = (\mathbf{A} - \lambda \mathbf{I})\mathbf{v}e^{\lambda t}.$$

Here, $e^{\lambda t}$ is never zero, so either $\mathbf{v} = 0$ or $(\mathbf{A} - \lambda \mathbf{I}) = 0$. We will assume $\mathbf{v} = 0$ and see what we obtain. If $(\mathbf{A} - \lambda \mathbf{I})\mathbf{v} = 0$ with $\mathbf{v}$ a nonzero vector, then the matrix $(\mathbf{A} - \lambda \mathbf{I})$ is singular and we must have

$$\det(\mathbf{A} - \lambda \mathbf{I}) = 0.$$

Solving the above equation will give us the eigenvalues λ, and using the relation

$$\mathbf{A}\mathbf{v} = \lambda \mathbf{v},$$

the corresponding eigenvectors can be obtained.

Algorithm

(1) Convert a DE of second order to a system of two first order DEs by introducing new variables.

(2) Once the system is set up, it has the form $\mathbf{x}' = \mathbf{A}\mathbf{x}$.

---

(3) Find the solutions through the use of the eigenvalues and eigenvectors.

(a) Solve $\det(\mathbf{A} - \lambda\mathbf{I}) = 0$ for λ to find the eigenvalues.

(b) Solve $\mathbf{Av} = \lambda\mathbf{v}$ to find the eigenvectors.

Remark

Eigenvalues could be (1) real and distinct, (2) real and equal, and (3) complex.

Real and distinct The eigenvectors found in step (3)-(b) are linearly independent and we can use the algorithm of Chapter 3 to find the solution.

Real and equal In this case, step (3) changes. If λ_1 is the repeated eigenvalue, then one solution to the system is

$$\mathbf{x}_1 = \mathbf{v}_1 e^{\lambda_1 t},$$

where $\mathbf{v}_1$ comes from $\mathbf{Av}_1 = \lambda_1\mathbf{v}_1$. However, the other linearly independent solution will be

$$\mathbf{x}_2 = te^{\lambda_1 t}\mathbf{v}_1 + e^{\lambda_1 t}\mathbf{v}_2,$$

where $\mathbf{v}_2$ comes from solving $(\mathbf{A} - \lambda_1\mathbf{I})\mathbf{v}_2 = \mathbf{v}_1$.

4.2.1 Real and Distinct Eigenvalues

Let's look at an example where we encounter real and distinct eigenvalues.

Example 4.2

Solve the IVP

$$2y'' + 5y' - 3y = 0, \qquad y(0) = -1, \ y'(0) = 4.$$

Solution:

Step 1 We set up the system:

$$\left.\begin{array}{rcl} x_1 & = & y \\ x_2 & = & y' \end{array}\right\} \quad \Rightarrow \quad \left.\begin{array}{rcl} x_1' & = & y' = x_2 \\ x_2' & = & y'' = -\dfrac{5}{2}y' + \dfrac{3}{2}y = -\dfrac{5}{2}x_2 + \dfrac{3}{2}x_1 \end{array}\right\}$$

We can write $\mathbf{x} = \left(\begin{smallmatrix} x_1 \\ x_2 \end{smallmatrix}\right) = \left(\begin{smallmatrix} y \\ y' \end{smallmatrix}\right)$. Also, since $y(0) = -1$ and $y'(0) = 4$, we have $\mathbf{x}(0) = \left(\begin{smallmatrix} -1 \\ 4 \end{smallmatrix}\right)$.

Step 2 The system, written in matrix form, is

$$\begin{pmatrix} x_1 \\ x_2 \end{pmatrix}' = \begin{pmatrix} 0 & 1 \\ \frac{3}{2} & -\frac{5}{2} \end{pmatrix} \begin{pmatrix} x_1 \\ x_2 \end{pmatrix}, \qquad \begin{pmatrix} x_1(0) \\ x_2(0) \end{pmatrix} = \begin{pmatrix} -1 \\ 4 \end{pmatrix}.$$

Step 3 Now we find the eigenvalues and eigenvectors:

$$\det \begin{pmatrix} 0 - \lambda & 1 \\ \frac{3}{2} & -\frac{5}{2} - \lambda \end{pmatrix} = 0,$$

that is,

$$-\lambda \left(-\frac{5}{3} - \lambda \right) - 1 \left(\frac{3}{2} \right) = 0.$$

Simplifying, this expression becomes

$$2\lambda^2 + 5\lambda - 3 = 0 \qquad \Rightarrow \qquad (2\lambda - 1)(\lambda + 3) = 0.$$

Hence, the eigenvalues are

$$\lambda_1 = \frac{1}{2} \qquad \text{and} \qquad \lambda_2 = -3,$$

so we have two real and distinct eigenvalues. Now we need to find the corresponding eigenvectors. Let's find the eigenvector for $\lambda_1 = \frac{1}{2}$. We have to solve the system

$$(\mathbf{A} - \lambda_1 \mathbf{I})\mathbf{v} = 0,$$

that is,

$$\begin{pmatrix} -\frac{1}{2} & 1 \\ \frac{3}{2} & -3 \end{pmatrix} \begin{pmatrix} v_1 \\ v_2 \end{pmatrix} = \begin{pmatrix} 0 \\ 0 \end{pmatrix}.$$

Solving this system, we find that the eigenvector corresponding to the eigenvalue λ_1 is

$$\mathbf{v}_1 = \begin{pmatrix} 1 \\ \frac{1}{2} \end{pmatrix}.$$

You can verify that the eigenvector corresponding to $\lambda_2 = -3$ is

$$\mathbf{v}_2 = \begin{pmatrix} 1 \\ -3 \end{pmatrix}.$$

Step 4 Now that we have found the eigenvalues and the corresponding eigenvectors, we can write the general solution:

$$\mathbf{x} = \begin{pmatrix} x_1 \\ x_2 \end{pmatrix} = C_1 e^{-3t} \begin{pmatrix} 1 \\ -3 \end{pmatrix} + C_2 e^{\frac{1}{2}t} \begin{pmatrix} 1 \\ \frac{1}{2} \end{pmatrix},$$

that is,

$$x_1(t) = C_1 e^{-3t}(1) + C_2 e^{\frac{1}{2}t}(1) = C_1 e^{-3t} + C_2 e^{\frac{1}{2}t}$$

and

$$x_2(t) = C_1 e^{-3t}(-3) + C_2 e^{\frac{1}{2}t} \left(\frac{1}{2} \right) = -3C_1 e^{-3t} + \frac{1}{2} C_2 e^{\frac{1}{2}t}.$$

Example 4.3

Solve the DE of example 4.2 using *Mathematica*.

Solution: In example 4.2 we set up the system

$$\begin{aligned} x_1' &= x_2 \\ x_2' &= \frac{3}{2}x_1 - \frac{5}{2}x_2 \end{aligned}$$

Let's define the coefficient matrix of this system in *Mathematica*:

In[1]:= `A = {{0, 1}, {`$\frac{3}{2}$`, -`$\frac{5}{2}$`}}`

Out[1]:= $\left\{\{0, 1\}, \{\frac{3}{2}, -\frac{5}{2}\}\right\}$

To display `A` as a matrix in conventional notation, we use the function **MatrixForm**:

In[2]:= `A`

Out[2]:= $\begin{pmatrix} 0 & 1 \\ \frac{3}{2} & -\frac{5}{2} \end{pmatrix}$

Now let's ask *Mathematica* to compute the eigenvalues and eigenvectors:

In[3]:= `evals = Eigenvalues[A]`

Out[3]:= $\left\{-3, \frac{1}{2}\right\}$

In[4]:= `evecs = Eigenvectors[A]`

Out[4]:= $\left\{\{-\frac{1}{3}, 1\}, \{2, 1\}\right\}$

Now we have all we need to construct the solution. Here is one way to do all at once in *Mathematica*:

In[5]:= `sol = Plus @@ MapThread[Times, {`$e^{\text{evals } t}$`, evecs}] /.`
　　　`{a_ + b_ :> C[1] a + C[2] b}`

Out[5]:= $\left\{-\frac{1}{3} e^{-3t} C[1] + 2 e^{t/2} C[2], \; e^{-3t} C[1] + e^{t/2} C[2]\right\}$

We know that the solution to our second order DE is the first component of this solution to the associated system. Hence, the solution is

In[6]:= `sol[[1]]`

Out[6]:= $-\frac{1}{3} e^{-3t} C[1] + 2 e^{t/2} C[2]$

4.2.2 Real and Equal Eigenvalues

Example 4.4

Solve the DE

$$y'' - 2y' + y = 0$$

Solution: First, we set up the system of DEs. We define

$$x_1(t) = y(t)$$
$$x_2(t) = y'(t)$$

so

$$x_1' = y' = x_2$$
$$x_2' = y'' = 2y' - y = 2x_2 - x_1$$

We can write this system using matrix notation:

$$\begin{pmatrix} x_1 \\ x_2 \end{pmatrix} = \begin{pmatrix} 0 & 1 \\ -1 & 2 \end{pmatrix} \begin{pmatrix} x_1 \\ x_2 \end{pmatrix}$$

so if we define $\mathbf{x} = \begin{pmatrix} x_1 \\ x_2 \end{pmatrix}$ and $\mathbf{A} = \begin{pmatrix} 0 & 1 \\ -1 & 2 \end{pmatrix}$, the system may be represented by

$$\mathbf{x}' = \mathbf{A}\mathbf{x}$$

Now we need to find the eigenvalues. How do we do this? we solve

$$det(\mathbf{A} - \lambda\mathbf{I}) = 0$$

for λ . We have that

$$det(\mathbf{A} - \lambda\mathbf{I}) = det \begin{pmatrix} 0 - \lambda & 1 \\ -1 & 2 - \lambda \end{pmatrix} = -\lambda(2 - \lambda) - (-1)(1) = \lambda^2 - 2\lambda + 1.$$

so to find the eigenvalues we solve

$$\lambda^2 - 2\lambda + 1 = 0 \Rightarrow (\lambda - 1)^2 = 0$$

so $\lambda = 1, 1$ is the repeated eigenvalue.

Now we find the eigenvectors. In the case of repeated eigenvalues, the eigenvector equation is different. For the repeated eigenvalue $\lambda = 1$, the first eigenvector equation is

$$(\mathbf{A} - \lambda\mathbf{I}) \begin{pmatrix} x_1 \\ x_2 \end{pmatrix} = \begin{pmatrix} 0 \\ 0 \end{pmatrix}$$

and for the second eigenvector, the equation we need to solve is

$$(\mathbf{A} - \lambda\mathbf{I})\begin{pmatrix} P_1 \\ P_2 \end{pmatrix} = \begin{pmatrix} x_1 \\ x_2 \end{pmatrix}$$

where given that $\lambda = 1$,

$$(\mathbf{A} - \lambda\mathbf{I}) = \begin{pmatrix} -1 & 1 \\ -1 & 1 \end{pmatrix}$$

Let's find the first eigenvector. Doing the row operation $R_2 \to R_2 - R_1$ on this matrix, we get

$$\begin{pmatrix} -1 & 1 \\ 0 & 0 \end{pmatrix}\begin{pmatrix} x_1 \\ x_2 \end{pmatrix} = \begin{pmatrix} 0 \\ 0 \end{pmatrix}$$

so $x_1 - x_2 = 0$, that is, $x_1 = x_2$. Therefore, out first eigenvector is

$$\mathbf{v_1} = \begin{pmatrix} 1 \\ 1 \end{pmatrix}$$

Now let's find the second eigenvector. We need to solve

$$\begin{pmatrix} -1 & 1 \\ -1 & 1 \end{pmatrix}\begin{pmatrix} P_1 \\ P_2 \end{pmatrix} = \begin{pmatrix} 1 \\ 1 \end{pmatrix}$$

Doing the row operation $R_2 \to R_2 - R_1$, we have

$$\begin{pmatrix} -1 & 1 \\ 0 & 0 \end{pmatrix}\begin{pmatrix} P_1 \\ P_2 \end{pmatrix} = \begin{pmatrix} 1 \\ 0 \end{pmatrix}$$

This implies that $-P_1 + P_2 = 1$

$$\Rightarrow P_2 = P_1 + 1$$
$$\Rightarrow P_1 = 1, P_2 = 2$$
$$\Rightarrow \mathbf{v_2} = \begin{pmatrix} 1 \\ 2 \end{pmatrix}$$

Note: Please check that $\mathbf{v_1} = \mathbf{v_2}$ are linearly independent.
Now we write the solution. Using the ideas of Chapter 3, we know that the solution to our DE is

$$y(t) = (C_1 + C_2 t)e^{\lambda t}$$

Note: Recall that we have defined $\begin{pmatrix} y \\ y' \end{pmatrix} = \begin{pmatrix} x_1 \\ x_2 \end{pmatrix}$. By analogy with the solution discussed in Chapter 3, we can write

$$\begin{pmatrix} y \\ y' \end{pmatrix} = (C_1 + C_2 t)e^{\lambda t}.$$

so,

$$\begin{pmatrix} y \\ y' \end{pmatrix} = \left(C_1\begin{pmatrix} 1 \\ 1 \end{pmatrix} + C_2\begin{pmatrix} 1 \\ 2 \end{pmatrix} t\right)e^t$$

$$= e^t\begin{pmatrix} C_1 \\ C_1 \end{pmatrix} + e^t t\begin{pmatrix} C_2 \\ 2C_2 \end{pmatrix}$$

Therefore, the solution to our DE is

$$y(t) = e^t + e^t 2C_2 t.$$

4.2.3 Complex Eigenvalues

Since systems of DEs makes us work with matrix system $\mathbf{x}' = \mathbf{A}\mathbf{x}$, the solution will be of the form $\mathbf{x} = \mathbf{v}e^{\lambda t}$ as discussed before.

Note:

(1) Complex eigenvalues of $\mathbf{A}$ will yield complex eigenvectors. The process to get rid of complex numbers is similar to the one executed on Chapter 3 (second order homogeneous differential equations).

(2) We are going to solve a second order homogeneous DE in two ways. The first method is the repetition of what was done in Chapter 3. The second method will be solving the matrix system. A comparative study will help us note a few important things.

Example 4.5

Solve the IVP

$$y'' - 2y' + 10y = 0, \qquad y(0) = 0,\ y'(0) = 1. \tag{4.7}$$

Solution:
Method 1: Using Chapter 3 discussion on solving a second order homogeneous DE.

(1) Assume $y = e^{mt}$ to be a trial solution to (4.7). Then the auxiliary equation is

$$m^2 - 2m + 10 = 0 \qquad \text{(a quadratic equation in } m\text{)}$$

(2) Therefore,

$$m = \frac{2 \pm \sqrt{4 - 40}}{2} = \frac{2 \pm \sqrt{-36}}{2} = \frac{2 \pm 6i}{2} = 1 \pm 3i.$$

(3) The solution is

$$\begin{aligned}
y(t) &= c_1 y_1 + c_1 y_2 \\
&= c_1 e^{(1+3i)t} + c_2 e^{(1-3i)t} \\
&= e^t \left[c_1 e^{3it} + c_2 e^{-3it} \right].
\end{aligned}$$

Using Euler's identity, $e^{i\theta} = \cos\theta + i\sin\theta$, the solution is rewritten as

$$y(t) = e^t(c_1 \cos 3t + c_2 \sin 3t),$$

where c_1 and c_2 are arbitrary constants of integration. Applying the initial condition, you can easily check that the solution to our IVP is

$$y(x) = \frac{1}{3}e^t \sin 3t.$$

Method 2: Using a system of DEs.

(1) First, we set up the system.

    (a) Introduce new variables:

$$x_1(t) = y(t) \tag{4.8}$$
$$x_2(t) = y'(t). \tag{4.9}$$

    (b) Differentiate with respect to t:

$$x_1'(t) = y'(t) = x_2 \qquad \text{(by (4.8))}$$

and

$$x_2'(t) = y''(t) = 2y' - 10y$$
$$= 2x_2 - 10x_1 \qquad \text{(by (4.9))}$$

    (c) The system of first order DEs is

$$x_1'(t) = x_2(t)$$
$$x_2'(t) = 2x_2 - 10x_1.$$

Using matrix notation, the above can be rewritten as

$$\begin{pmatrix} x_1 \\ x_2 \end{pmatrix}' = \begin{pmatrix} 0 & 1 \\ -10 & 2 \end{pmatrix}\begin{pmatrix} x_1 \\ x_2 \end{pmatrix}.$$

Please note that

$$\begin{pmatrix} x_1(t) \\ x_2(t) \end{pmatrix} = \begin{pmatrix} y(t) \\ y'(t) \end{pmatrix},$$

so

$$\begin{pmatrix} x_1(0) \\ x_2(0) \end{pmatrix} = \begin{pmatrix} y(0) \\ y'(0) \end{pmatrix} = \begin{pmatrix} 0 \\ 1 \end{pmatrix}.$$

    (d) The final form of our system is

$$\begin{pmatrix} x_1 \\ x_2 \end{pmatrix}' = \begin{pmatrix} 0 & 1 \\ -10 & 2 \end{pmatrix}\begin{pmatrix} x_1 \\ x_2 \end{pmatrix}, \qquad \begin{pmatrix} x_1(0) \\ x_2(0) \end{pmatrix} = \begin{pmatrix} 0 \\ 1 \end{pmatrix},$$

and we will define

$$\mathbf{A} = \begin{pmatrix} 0 & 1 \\ -10 & 2 \end{pmatrix}.$$

Remark: Eigenvalues and eigenvectors of $\mathbf{A}$ describe the solution of the system.

Compute eigenvalues and eigenvectors of $\mathbf{A}$:

Evaluate

$$
\begin{aligned}
\det(\mathbf{A} - \lambda\mathbf{I}) = \det \begin{pmatrix} 0 - \lambda & 1 \\ -10 & 2 - \lambda \end{pmatrix} &= \det \begin{pmatrix} -\lambda & 1 \\ -10 & 2 - \lambda \end{pmatrix}. \\
&= (-\lambda)(2 - \lambda) + 10 \\
&= -2\lambda + \lambda^2 + 10 \\
&= \lambda^2 - 2\lambda + 10.
\end{aligned}
$$

Set $\det(\mathbf{A} - \lambda\mathbf{I}) = 0$, i.e.

$$\lambda^2 - 2\lambda + 10 = 0.$$

Let's stop here to notice something. The auxiliary equation in step 1 of method 1 in this example is

$$m^2 - 2m + 10 = 0$$

and the eigenvalue equation is $\lambda^2 - 2\lambda + 10 = 0$. This is not a coincidence. In fact, every system of DEs corresponding to a second order homogeneous DE will exhibit this property, i.e. the auxiliary equation (in this case $m^2 - 2m + 10 = 0$) and the characteristic equation (in this case $\lambda^2 - 2\lambda + 10 = 0$) are basically the same, except that in the characteristic equation the letter m is replaced by λ.

How is this observation helpful? Since

$$m = 1 \pm 3i \qquad \Rightarrow \qquad \lambda = 1 \pm 3i,$$

if you want to avoid evaluating the determinant, you have a choice.

(2) Ok, so let's continue with the next step, namely, computing the eigenvectors. We need to solve the system

$$(\mathbf{A} - \lambda\mathbf{I}) \begin{pmatrix} v_1 \\ v_2 \end{pmatrix} = \begin{pmatrix} 0 \\ 0 \end{pmatrix}.$$

In this case,

$$\mathbf{A} = \begin{pmatrix} -\lambda & 1 \\ -10 & 2 - \lambda \end{pmatrix},$$

so for $\lambda = 1 + 3i$, we need to solve the system

$$\begin{pmatrix} -1 - 3i & 1 \\ -10 & 1 - 3i \end{pmatrix} \begin{pmatrix} v_1 \\ v_2 \end{pmatrix} = \begin{pmatrix} 0 \\ 0 \end{pmatrix}.$$

Remark: In order to solve for v_1 and v_2, please note that we are seeking a relation between v_1 and v_2, not a specific solution. Therefore, you are at liberty to use any one row. Using the first row, we have

$$(-1 - 3i)v_1 + v_2 = 0,$$

so
$$v_2 = (1 + 3i)v_1.$$

So, the first eigenvector is

$$\mathbf{v}_1 = \begin{pmatrix} v_1 \\ (1+3i)v_1 \end{pmatrix}.$$

Choosing $v_1 = 1$, we get $\mathbf{v}_1 = \begin{pmatrix} 1 \\ (1+3i) \end{pmatrix}$ (make a reasonable choice every time, i.e. in this case if you choose $v_1 = 0$ you get $\mathbf{v}_1 = \begin{pmatrix} 0 \\ 0 \end{pmatrix}$, which doesn't make sense.) **Remark:** Complex numbers should appear in the numerator of any fraction while trying to find the eigenvectors. For example, if

$$v_2 = \frac{1}{1 - 3i},$$

rewrite v_2 as

$$\frac{1 + 3i}{10}.$$

To find the second eigenvector, we solve

$$\begin{pmatrix} -1+3i & 1 \\ -10 & 1+3i \end{pmatrix} \begin{pmatrix} v_1 \\ v_2 \end{pmatrix} = \begin{pmatrix} 0 \\ 0 \end{pmatrix}.$$

Once again, using the first row, we have

$$(-1 + 3i)v_1 + v_2 = 0,$$

so

$$v_2 = (1 - 3i)v_1,$$

and the second eigenvector is

$$\mathbf{v}_2 = \begin{pmatrix} v_1 \\ (1-3i)v_1 \end{pmatrix}.$$

Choosing $v_1 = 1$, we get $\mathbf{v}_2 = \begin{pmatrix} 1 \\ 1-3i \end{pmatrix}$.

(3) The form of the solution is

$$\begin{pmatrix} x_1(t) \\ x_2(t) \end{pmatrix} = c_1 e^{(1+3i)t} \begin{pmatrix} 1 \\ 1+3i \end{pmatrix} + c_2 e^{(1-3i)t} \begin{pmatrix} 1 \\ 1-3i \end{pmatrix},$$

that is,

$$\begin{pmatrix} x_1(t) \\ x_2(t) \end{pmatrix} = \begin{pmatrix} c_1 e^{(1+3i)t} + c_2 e^{(1-3i)t} \\ (1+3i)c_1 e^{(1+3i)t} + (1-3i)c_2 e^{(1-3i)t} \end{pmatrix}.$$

Recall that our task is to find $y(t)$, which is by construction $x_1(t)$. Therefore, we read the first column of the above matrix, and find that

$$y(t) = x_1(t) = c_1 e^{(1+3i)t} + c_2 e^{(1-3i)t}. \tag{4.10}$$

(4) Now the final step: the following tasks need to be performed.

(a) Compute c_1 and c_2 applying the initial condition

$$\begin{pmatrix} x_1(0) \\ x_2(0) \end{pmatrix} = \begin{pmatrix} 0 \\ 1 \end{pmatrix}.$$

(b) Get rid of complex numbers from the expression defining $y(t)$.

Let's start by computing c_1 and c_2.

$$\begin{pmatrix} x_1(t) \\ x_2(t) \end{pmatrix} = \begin{pmatrix} c_1 e^{(1+3i)t} + c_2 e^{(1-3i)t} \\ (1+3i)c_1 e^{(1+3i)t} + (1-3i)c_2 e^{(1-3i)t} \end{pmatrix},$$

so for $\begin{pmatrix} x_1(0) \\ x_2(0) \end{pmatrix} = \begin{pmatrix} 0 \\ 1 \end{pmatrix}$, we get

$$\begin{pmatrix} 0 \\ 1 \end{pmatrix} = \begin{pmatrix} c_1 + c_2 \\ c_1(1+3i) + c_2(1-3i) \end{pmatrix}.$$

This implies

$$c_1 + c_2 = 0$$

and

$$3(c_1 - c_2)i = 1 \quad \Rightarrow \quad (c_1 - c_2)i = \frac{1}{3}.$$

From **(4.10)**, we have

$$\begin{aligned} x_1(t) = y(t) &= c_1 e^{(1+3i)t} + c_2 e^{(1-3i)t} \\ &= c_1 e^t e^{3it} + c_2 e^t e^{-3it} \\ &= e^t \left[c_1 e^{3it} + c_2 e^{-3it} \right] \\ &= e^t \left[c_1 \cos 3t + i c_2 \sin 3t + c_2 \cos 3t - i c_1 \sin 3t \right] \\ &= e^t \left[(c_1 + c_2) \cos 3t + i(c_1 - c_2) \sin 3t \right]. \end{aligned}$$

Using the results $c_1 + c_2 = 0$ and $i(c_1 - c_2) = \frac{1}{3}$, we get

$$y(t) = \frac{1}{3} e^t \sin 3t,$$

which is the solution obtained by using method 1 as well.

Concluding note: Agreed, this is a tedious method; but the method itself is not different from the case where the eigenvalues and eigenvectors are real. The extra work involved is the algebra of complex numbers and the ability to use Euler's identity efficiently.

Exercises

(1) Set up the following DEs as a matrix system of first order DEs.

    (a) $y'' - 7y' + 10y = 0$.

    (b) $y'' - 11y' + 28y = 0$.

    (c) $y'' - 27y = 0$.

    (d) $3y''' - 4y'' + y' - 8y = 0$.

    (e) $y'' - 2y' + y = 0$.

    (f) $y'' + 3y = 0$.

(2) Now solve (1), (3), (5) and (6) completely by using the system you have set up already.

Chapter 5

Qualitative Techniques

In this chapter we will look at qualitative techniques to study differential equations. The reason is that, as we have mentioned already, there are many differential equations that cannot be solved explicitly. However, in many cases we are not concerned with actually finding an explicit solution to a differential equation, but in a particular characteristic of the nature of the solution. For example, we may be interested in knowing if a small error in our measurement of initial conditions could cause our solution to be "very different" from the real solution that would be obtained if our measurements were completely accurate. To answer this, we will look at the property of **stability** or **unstability** of a solution. We will start with a qualitative analysis of first order autonomous DE, and then we will extend our discussion to the qualitative analysis of systems of first order autonomous differential equations.

5.1   The Phase Portrait

In this section, we will begin our qualitative analysis of first order autonomous DEs, that is, differential equations of the form

$$\frac{dy}{dx} = f(y), \tag{5.1}$$

where the right-hand side does not involve the independent variable x. We will start by introducing some terminology.

A **critical point** of **(5.1)** is a zero of $f(x)$, that is, a number c such that $f(c) = 0$. If c is a critical point of **(5.1)**, then $y = c$ is a constant solution to this DE. We will refer to constant solutions as **equilibrium solutions**.

Our qualitative analysis will merely consist of classifying critical points of an autonomous DE. The information needed for such classification can be reflected in a diagram called the **phase portrait** of a differential equation. To see what this is, let's look at an example.

Example 5.1

Draw the phase portrait of the DE

$$y' = y^2 - 3y + 2.$$

Solution: First, let's find the critical points of the DE. Since

$$f(y) = y^2 - 3y + 2 = (y-1)(y-2),$$

we see that 1 and 2 are critical points of the DE, so $y = 1$ and $y = 2$ are equilibrium solutions to the DE, and these are the unique solutions to the DE when the initial conditions are $y(0) = 1$ and $y(0) = 2$, respectively. To investigate the qualitative behaviour of solutions other than these equilibrium solutions, let's look at the sign of $y' = f(y) = (y-1)(y-2)$ for values of y in the intervals

$$(-\infty, 1), \qquad (1, 2), \qquad (2, \infty),$$

which will give us information on the increasing and decreasing behavior of the solutions. Suppose we have the initial condition $y(0) = y_0$. If $y_0 \in (-\infty, 1)$, then both factors $(y-1)$ and $(y-2)$ will be negative, so $y' = (y-1)(y-2) > 0$. This means that any solution to the DE with the initial condition $y_0 \in (-\infty, 1)$ will be increasing for all x, since the uniqueness and existence theorem guarantees $y \in (-\infty, 1)$ for all x. On the other hand, if $y_0 \in (1, 2)$, the solution will be decreasing, for in this case $y' = (y-1)(y-2) < 0$. Finally, if $y_0 \in (2, \infty)$, then the solution to the DE will be increasing. This information can be summarized in the following diagram:

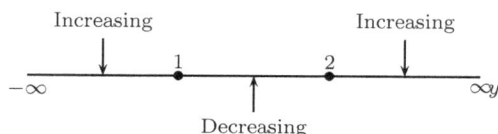

Instead of the labels "increasing" and "decreasing", we can draw arrows pointing to the right or to the left depending on whether the solution is increasing or decreasing, respectively:

The diagram above is the **phase portrait** for the differential equation $y' = y^2 - 3y + 2 = (y-1)(y-2)$. Let's look at the direction field for this differential equation to see its relation to the phase portrait.

```
In[1]:= Clear[dField]; dField =
    VectorPlot[Normalize[{1, (y - 1) (y - 2)}], {x, 0, 5}, {y, -2, 4},
      VectorScale → 0.03, VectorPoints → 19, VectorStyle → Gray,
      Epilog → {PointSize[Medium], Line[{{0, -2}, {0, 4}}], Point /@
        {{0, 1}, {0, 2}}, Arrowheads[0.07], Arrow[{{0, 2}, {0, 3}}],
        Arrow[{{0, 2}, {0, 1.3}}], Arrow[{{0, 1}, {0, 0}}]}]
```

Out[1]:=

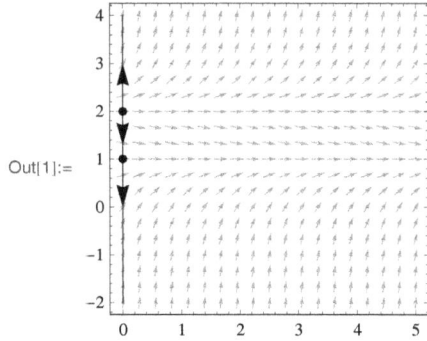

We have included the phase portrait in the direction field to make their relation more evident. The direction field shows how the solutions to the DE with initial condition $y(0) \in (-\infty, 1)$ are increasing and approach the equilibrium solution $y \equiv 1$. When the initial condition satisfies $y(0) \in (1, 2)$, the solutions are decreasing and also approach the equilibrium solution $y \equiv 1$. On the other hand, when the initial condition satisfies $y(0) \in (2, \infty)$, we see that the solutions are increasing and expedited away from the equilibrium solution $y \equiv 2$. Let's make this even more evident by drawing approximations to some integral curves for the solutions with initial conditions

$$y(0) = -1, 0, 1, 2, 1.5, 2.01, 2.1, 3.$$

For this, we will use the *Mathematica* function **NDSolve**, which is used to numerically solve a differential equation. Of course, our DE in this example can be solved explicitly, but that is something we are not interested in this time, for we want to investigate the nature of the solutions and to develop the techniques that will allow us to do this, and that can be applied to other, more complex differential equations.

To have *Mathematica* find numerical solutions to our DE for the different initial conditions mentioned above, we will use the function **Table** to repeatedly evaluate the function **NDSolve**, each time with a different value. Here is how we can do it:

```
In[2]:= Clear[sols, intCurves];
        sols = Table[NDSolve[{y'[x] == (y[x] - 1) (y[x] - 2), y[0] == k},
          y[x], {x, 0, 4}], {k, {-1, 0, 1, 2, 1.5, 2.01, 2.1, 3}}];
        intCurves = Plot[Evaluate[y[x] /. sols], {x, 0, 5},
          PlotRange -> {-2, 4}, PlotStyle -> Thick];
        Show[{dField, intCurves}, AspectRatio -> Automatic]
```

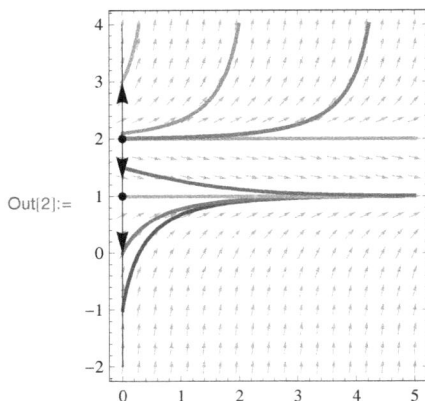

Out[2]:=

Note that the solutions that satisfy initial conditions $-\infty < y(0) < 1$ approach $y \equiv 1$ as $t \to \infty$. The same is true for solution corresponding to initial conditions in the interval $1 < y(0) < 2$. We can therefore say that initial conditions near the critical point $y = 1$ will remain close to $y = 1$ and, indeed, converge to this critical point as $t \to \infty$. This being the case, we say that the critical point $y = 1$ is a **sink**.

The situation with the critical point $y = 2$ is different. Note that any solution that is not the equilibrium $y \equiv 2$ moves away from $y \equiv 2$ as $t \to \infty$, and this is true no matter how close the initial condition is to the critical point $y = 2$ (as long as $y(0)$ is not 2 itself). This property of the critical point $y = 2$ is the opposite of that of $y = 1$, and in this case we say that the critical point $y = 2$ is a **source**.

Consider again the equilibrium solution $y \equiv 1$. We have seen that any solution to the DE with an initial condition that is close to 1 will remain close to 1. For this reason, we say that such solution is **stable**.

5.2   The Phase Plane of a System of DEs

Now we will look at the qualitative study of the solutions of an autonomous system of DEs. Let's introduce the terminology of this section.

First, we define an **autonomous system** of two first order differential equations as a system of the form

$$\begin{cases} \frac{dx}{dt} & = & f(x,y) \\ \frac{dy}{dt} & = & g(x,y) \end{cases}$$

Therefore, in the differential equations of an autonomous system, the independent variable t does not appear explicitly. If

$$\begin{pmatrix} x(t) \\ y(t) \end{pmatrix}$$

is a solution to the above system, then this vector describes a parametrized solution curve in the xy-plane, which we call the **phase plane**. We say that a point (x^*, y^*) of the phase plane is a **critical point** of the system if

$$f(x^*, y^*) = 0 \qquad \text{and} \qquad g(x^*, y^*) = 0.$$

If (x^*, y^*) is a critical point of the system, then

$$\mathbf{Y}(y) \equiv \begin{pmatrix} x^* \\ y^* \end{pmatrix}$$

is an **equilibrium solution** (a constant solution) to the system. The **phase portrait** of the system will then consist of the graph of typical solution curves to the system, including all of its equilibrium solutions.

Example 5.2

Draw a phase portrait for the system

$$\begin{aligned} \frac{dx}{dt} & = & -2x + 3y \\ \frac{dy}{dt} & = & -x - 2y \end{aligned}$$

Solution: The critical points of this system are found by solving the system of linear equations

$$\begin{aligned} -2x + 3y & = & 0 \\ -x - 2y & = & 0. \end{aligned}$$

The only solution to this system is the trivial solution $(0, 0)$. therefore, the only critical point of the system is $(0, 0)$. We can solve this system with the analytical techniques discussed in the previous chapter, but instead we will find approximate solutions using the *Mathematica* function **NDSolve**. But first, let's look at the vector field of this system.

```
In[1]:=  Clear[x, y, dField];
         dField = VectorPlot[{-2 x + 3 y, -x - 2 y}, {x, -5, 5}, {y, -5, 5}]
```

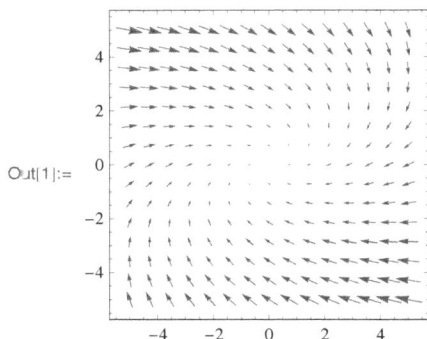

Out[1]:=

The arrows of the direction field obtained above represent the vectors of the
direction field of the vector function

$$\mathbf{r}(t) = \begin{pmatrix} -2x + 3y \\ -x - 2y \end{pmatrix}.$$

Since the vectors of the vector field have different magnitudes at different
points, the arrows are drawn with different lengths. If we want all of the
arrows of our direction field to have the same length, we can use the function
Normalize to make all vectors unit vectors and the option **VectorScale** to
specify a length for all arrows. Here is the difference (option **VectorPoints** is
used to plot $19 \times 19 = 361$ arrows):

```
In[2]:=  dField = VectorPlot[Normalize[{-2 x + 3 y, -x - 2 y}], {x, -5, 5},
           {y, -5, 5}, VectorScale → 0.03, VectorPoints → 19]
```

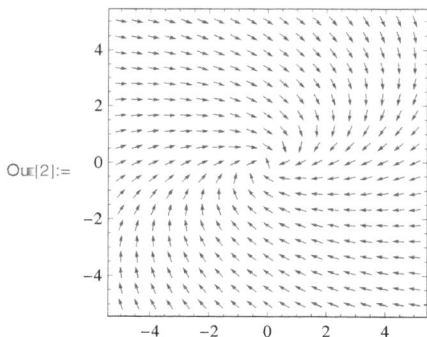

Out[2]:=

Note that the direction field gives us enough information to figure out the
behavior of the solutions to our system of differential equations. The arrows
indicate that the solutions to the system with initial conditions anywhere in the
region $[-5, 5] \times [-5, 5]$ will approach $(0, 0)$ as $t \to \infty$ (the arrows point in the
direction of increasing t). Let's plot some solution curves in the phase plane.

As we mentioned, these curves are defined by their corresponding solution vector

$$\begin{pmatrix} x(t) \\ y(t) \end{pmatrix},$$

in such a way that a point (a, b) is in the solution curve defined by this vector if $(a, b) = (x(t), y(t))$ for some t in the domain of the solution. To plot such curves, we can use *Mathematica* function **ParametricPlot**. So let's ask *Mathematica* for an approximate solution to the system for the initial conditions $x(0) = 5$ and $y(0) = 5$ and then to plot it with the function **ParametricPlot**. Here is how we can do it:

```
In[3]:= Clear[sol];
        sol = NDSolve[{x'[t] == -2 x[t] + 3 y[t], y'[t] == -x[t] - 2 y[t],
            x[0] == 5, y[0] == 5}, {x[t], y[t]}, {t, 0, 10}];
        ParametricPlot[{x[t], y[t]} /. sol, {t, 0, 10},
          PlotRange → {{-6, 6}, {-5, 5}}]
```

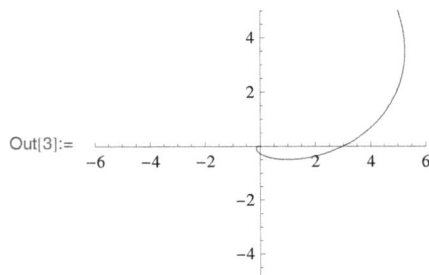

Out[3]:=

This solution curve agrees with our prediction: the curve approaches $(0, 0)$ as t increases. This is true for any solution curve regardless of the initial conditions given. For this reason, we say that the point $(0, 0)$ is a sink, and that any solution near this critical point is stable, since any solution curve $(0, 0)$ will remain close to $(0, 0)$. Figure 5.1 shows a *Mathematica* program, **phasePortrait**, that plots a phase portrait with solution curves corresponding to the specified initial conditions. We can use it to plot a phase portrait for the system of this and other examples.

In[4]:= **phasePortrait[{x′[t] == -2 x[t] + 3 y[t], y′[t] == -x[t] - 2 y[t]},**
     {x, -5, 5}, {y, -5, 5}, {{5, 5}, {-5, 5}, {5, -5}}, {t, 0, 3}]

Out[4]:=

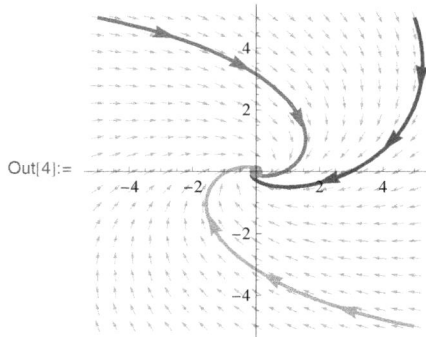

Let's look at another example of a system with stable solutions. This time, solutions near a critical point will remain close to this critical point but they will not approach it asymptotically as in the previous example.

Example 5.3

Draw a phase portrait for the system

$$\frac{dx}{dt} = y$$

$$\frac{dy}{dt} = -\frac{1}{2}x$$

Solution: This system is quite simple, but it will serve to illustrate a critical point that is called a **center**. We see right away that the only critical point of this system is the point $(0,0)$. We use our *Mathematica* function **phasePortrait**, defined in figure 5.1, to look at typical the vector field of this system and some of its typical solution curves.

In[1]:= **phasePortrait[{x′[t] == y[t], y′[t] == -x[t]}, {x, -5, 5},**
     {y, -5, 5}, {{4, 0}, {3, 0}, {2, 0}, {1, 0}}, {t, 0, 2 π}]

Out[1]:=

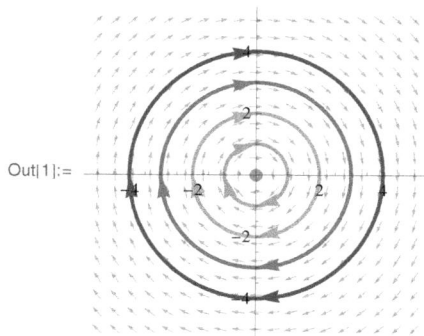

Note that solutions that are near the critical point $(0,0)$ always remain close to this point, since the solution curves only orbit around $(0,0)$. For this reason, we call the critical point $(0,0)$ a center.

```
phasePortrait[{eqns__}, {var1_, xmin_, xmax_},
  {var2_, ymin_, ymax_}, {iniCond__List},
  {t_, tmin_, tmax_}] :=
 Module[{sols, criticalPts, intCurves, dfield},
  sols =
   NDSolve[
     {eqns}~Join~{var1[0] == #[[1]],
        var2[0] == #[[2]]} // Flatten,
      {var1[t], var2[t]}, {t, tmin, tmax}] & /@ {iniCond};
  criticalPts =
   Solve[{eqns} /. {var1'[t] -> 0, var2'[t] -> 0},
    {var1[t], var2[t]}, InverseFunctions -> True];
  criticalPts = {var1[t], var2[t]} /. criticalPts;
  intCurves =
   ParametricPlot[Evaluate[{var1[t], var2[t]} /. sols],
     {t, tmin, tmax},
     PlotRange -> {{xmin, xmax}, {ymin, ymax}},
     PlotStyle ->
      {{Thick, Arrowheads[{{.07, .5}, {.07, .25},
           {.07, .75}}]}}] /. {Line -> Arrow};
  dfield = VectorPlot[Normalize[{eqns}[[All, 2]]],
    {var1[t], xmin, xmax}, {var2[t], ymin, ymax},
    VectorScale -> .03, VectorPoints -> 19,
    VectorStyle -> Gray];
  Show[{dfield, intCurves,
    Graphics[{PointSize[Large], Red,
      Point /@ criticalPts}]}, Frame -> False, Axes -> True]
 ]
```

Figure 5.1. A function to plot phase portraits of systems of differential equations.

Exercises

(1) Find the critical points of the system of the given autonomous system. Then use *Mathematica* to plot the direction field of the system, and then use the result to sketch a few solution curves by hand.

(a) $\dfrac{dx}{dt} = 2x - y, \quad \dfrac{dy}{dt} = x - 2y.$

(b) $\dfrac{dx}{dt} = 2x - 3y, \quad \dfrac{dy}{dt} = 4x + 2y - 1.$

(c) $\dfrac{dx}{dt} = xy - y^2, \quad \dfrac{dy}{dt} = xy + x^2.$

(d) $\dfrac{dx}{dt} = \cos y, \quad \dfrac{dy}{dt} = -x.$

(2) Below, the coefficient matrix $\mathbf{A}$ of the system $\mathbf{x}' = Ax$ is given. Using *Mathematica*, find the eigenvalues and eigenvectors of $\mathbf{A}$, and then find the solution to the system. Plot several solutions to each system using the function **ParametricPlot**, and relate what you see with the following possibilities: the eigenvalues of $\mathbf{A}$ are (1) real unequal and of the same sign, (2) real unequal and of opposite sign (3) real nonzero and equal, (4) complex conjugate, and (5) pure imaginary.

(a) $\mathbf{A} = \begin{pmatrix} -1 & 0 \\ 0 & -1 \end{pmatrix}.$

(b) $\mathbf{A} = \begin{pmatrix} 0 & 2 \\ -2 & 0 \end{pmatrix}.$

(c) $\mathbf{A} = \begin{pmatrix} 2 & 0 \\ 0 & 1 \end{pmatrix}.$

(d) $\mathbf{A} = \begin{pmatrix} 2 & 0 \\ 0 & -2 \end{pmatrix}.$

(e) $\mathbf{A} = \begin{pmatrix} 1 & -2 \\ 2 & 1 \end{pmatrix}.$

Chapter 6

The Laplace Transform Method

In order to understand the ideas of this chapter, you have to make sure you understand the ideas and result on the Laplace transform introduced in chapter 0, which you may consider necessary to revisit before starting this chapter.

In this chapter we will look at another method for solving IVPs like

$$ay'' + by' + cy = g(x), \qquad y(0) = k_1 \text{ and } y'(0) = k_2,$$

which is a second order, linear, non-homogeneous DE with constant coefficients with initial conditions $y(0) = k_1$ and $y'(0) = k_2$. Our new method will use the Laplace transform to turn our initial value problem into an algebraic problem, and it is particularly useful when dealing with cases in which the function $g(x)$ involves the unit step function or the impulse function, cases that we have not considered so far and that have important applications.

6.1 Solving IVPs with the Laplace Transform

6.1.1 Constant Coefficients DE

We will begin with an example. Note that the IVP of the example can be solved easily using the integrating factor method. Indeed, you will immediately realize that the integrating factor method gets you to the solution more easily than using the Laplace transform. Keep in mind, however, that the Laplace transform will be used to solve IVPs that involve functions we do not know how to deal with so far, and that the following example is there only to illustrate the Laplace transform method with a problem that does not require too much work, and hence it let's you concentrate on the crucial steps.

Example 6.1

Use the Laplace transform to solve the following IVP.

$$y' - 2y = x, \qquad y(0) = 3.$$

Solution: The first thing we do is to compute the Laplace transform of both sides of the DE. Because of the linearity property of the Laplace transform, taking the LT of each side in the equation is the same as taking the LT of each term, so that is what we will do:

$$\mathcal{L}[y'] - 2\mathcal{L}[y] = \mathcal{L}[x].$$

Now you have to look at the table of Laplace transforms and use the appropriate transformations for each term of the equation. Using this transforms, we have

$$sY(s) - y(0) - 2Y(s) = \frac{1}{s^2}.$$

Recall that $Y(s)$ is simply a short name for $\mathcal{L}[y](s)$, the Laplace transform of the function y. Now we substitute the given initial condition $y(0) = 3$ and factor $Y(s)$, which gives us

$$(s - 2)Y(s) - 3 = \frac{1}{s^2}.$$

Hence,

$$Y(s) = \frac{1}{s^2(s-2)} + \frac{3}{s-2} = \frac{1 + 3s^2}{s^2(s-2)}.$$

What we have here is the Laplace transform of our solution $Y(s)$. If we want to find the solution, we need to take the inverse LT of both sides of the equation. In other words, we need to figure out which is the function whose Laplace transform is the right-hand side of the above equation. However, the right-hand side is not written in a form that allows us to answer this question right away. Here we have to use partial fraction decomposition to write the right-hand side as the sum of simpler fractions, and hopefully we will be able to recognize each simpler fraction as the LT of some function. Here is the result of the partial fraction decomposition of the right-hand side of the equation

$$Y(s) = \frac{13}{4(s-2)} - \frac{1}{4s} - \frac{1}{2s^2}.$$

Can you recognize each term in the right-hand side as the LT of some function? Look at the table of LTs, and you will see that

$$\frac{13}{4(s-2)} = \frac{13}{4}\mathcal{L}[e^{2x}],$$

$$\frac{1}{4s} = \frac{1}{4}\mathcal{L}[1],$$

$$\frac{1}{2s^2} = \frac{1}{2}\mathcal{L}[x].$$

Therefore, we have

$$\mathcal{L}[y] = \frac{13}{4}\mathcal{L}\left[e^{2x}\right] - \frac{1}{4}\mathcal{L}[1] - \frac{1}{2}\mathcal{L}[x].$$

If the above equation is the LT of our solution, then the solution itself is

$$y = \frac{13}{4}e^{2x} - \frac{1}{4} - \frac{1}{2}x.$$

Before looking at another example, let's describe the general procedure for solving a DE using the Laplace transform.

Solving an IVP Using the Laplace Transform

(1) We compute the Laplace transform of both sides of the equation. Because of the linearity property of the transform, we actually compute the Laplace transform of each term.

(2) Computing the Laplace transform of the derivatives of our unknown function introduces initial values like $y(0)$, $y'(0)$, etc., so substitute these as indicated by the initial conditions of the IVP.

(3) Collect all the terms containing the Laplace transform of our function, say $Y(s)$. Then factor $Y(s)$ and isolate it in the equation.

(4) The right-hand side of our equation now contains an expression in terms of the variable s. Use partial fraction decomposition to write this expression as the sum of simpler fractions.

(5) Recognize each term in the right-hand side of the equation as the Laplace transform of some function.

(6) Apply the inverse of the Laplace transform to the equation to obtain the solution.

We will see some examples where the partial fraction decomposition in step (4) is avoided by applying the convolution theorem for the Laplace transform. To see how this is done, look at example 6.4

Now let's look at another example, this time involving a second order differential equation.

Example 6.2

Solve the IVP

$$y'' - 3y' + 2y = 2t, \qquad y(0) = -1,\ y'(0) = -1.$$

Solution: Again, we start by taking the Laplace transform of each term in the equation:

$$\mathcal{L}\left[y''\right] - 3\mathcal{L}\left[y'\right] + 2\mathcal{L}\left[y\right] = 2\mathcal{L}\left[t\right],$$

and using the table of Laplace transforms, we compute each of this transforms, obtaining

$$s^2 Y(s) - sy(0) - y'(0) - 3(sY(s) - y(0)) + 2Y(s) = \frac{2}{s^2}.$$

Applying the initial conditions, we have

$$s^2 Y(s) + s + 1 - 3(sY(s) + 1) + 2Y(s) = \frac{2}{s^2}.$$

We simplify and factor $Y(s)$ to obtain

$$(s^2 - 3s + 2)Y(s) + s - 2 = \frac{2}{s^2},$$

which means that

$$Y(s) = \frac{2}{s^2(s^2 - 3s + 2)} - \frac{s - 2}{s^2 - 3s + 2} = \frac{2 - s^3 + 2s^2}{s^2(s^2 - 3s + 2)},$$

that is,

$$Y(s) = \frac{2 - s^3 + 2s^2}{s^2(s - 2)(s - 1)}.$$

Now we need to use partial fraction decomposition. Here we will give you the result, but make sure you work out the decomposition yourself. We have

$$Y(s) = \frac{1}{s^2} + \frac{3}{2s} - \frac{3}{s - 1} + \frac{1}{2(s - 2)}.$$

Looking at the table of Laplace transforms, we see that

$$Y(s) = \frac{1}{2}\mathcal{L}\left[t\right] + \frac{3}{2}\mathcal{L}\left[1\right] - 3\mathcal{L}\left[e^t\right] + \frac{1}{2}\mathcal{L}\left[e^{2t}\right],$$

so finally the solution is

$$y(t) = t + \frac{3}{2} - 3e^t + \frac{1}{2}e^{2t}.$$

Now let's use *Mathematica* to solve the IVP of the last example.

Example 6.3

Use *Mathematica* to solve the IVP

$$y'' - 3y' + 2y = 2t, \qquad y(0) = -1, \ y'(0) = -1.$$

Solution: We use the *Mathematica* function

$$\texttt{LaplaceTransform[y,t,s]}$$

to compute the Laplace transform of a function y of t. First, let's define

In[1]:= `Clear[y, t, Y, s]; DE = (y')'[t] - 3 y'[t] + 2 y[t] == 2 t`

Out[1]:= $2\,y[t] - 3\,y'[t] + y''[t] == 2\,t$

Now we compute the Laplace transform of the equation:

In[2]:= `LT = LaplaceTransform[DE, t, s]`

Out[2]:= $2\,\texttt{LaplaceTransform}[y[t], t, s] + s^2\,\texttt{LaplaceTransform}[y[t], t, s] -$
$3\,(s\,\texttt{LaplaceTransform}[y[t], t, s] - y[0]) - s\,y[0] - y'[0] == \frac{2}{s^2}$

This result is better read if we replace each occurrence of **LaplaceTransform[y[t],t,s]** with **Y[s]**. We will also use a replacement instruction to substitute the initial conditions into the result:

In[3]:= `LT =`
    `LT /. {LaplaceTransform[y[t], t, s] -> Y[s], y[0] -> -1, y'[0] -> -1}`

Out[3]:= $1 + s + 2\,Y[s] + s^2\,Y[s] - 3\,(1 + s\,Y[s]) == \frac{2}{s^2}$

We can ask *Mathematica* to solve the above result for the function $Y(s)$. The *Mathematica* function **Solve** returns a rule that we can turn into an assignment. This is done so:

In[4]:= `sol = Solve[LT, Y[s]][[1, 1]];`
    `Set @@ sol`

Out[4]:= $\frac{2 + 2\,s^2 - s^3}{s^2\,(2 - 3\,s + s^2)}$

Now **Y[s]** has the following value:

In[5]:= `Y[s]`

Out[5]:= $\frac{2 + 2\,s^2 - s^3}{s^2\,(2 - 3\,s + s^2)}$

We can ask *Mathematica* to directly compute the inverse Laplace transform of this function to obtain the solution.

In[6]:= `InverseLaplaceTransform[Y[s], s, x]`

Out[6]:= $\frac{3}{2} - 3\,e^x + \frac{e^{2\,x}}{2} + x$

And that's it, we have solved the IVP by the Laplace transform method.

6.1.2   Using the Convolution Theorem

Recall that the convolution of two functions f and g is the function $f * g$ defined by

$$(f * g)(x) = \int_0^x f(\xi)g(x - \xi) \, d\xi,$$

provided this integral exists for $x > 0$. Our interest in this operation lies in the fact that

$$\mathcal{L}\left[f(x)\right] \mathcal{L}\left[g(x)\right] = \mathcal{L}\left[(f * g)(x)\right],$$

whenever the Laplace transform of f and g exist. This fact is called the **convolution theorem**, and we can use it to find the inverse Laplace transform of certain expressions without having to compute its partial fraction decomposition. Let's look at an example.

Example 6.4

Solve the IVP

$$y' - 2y = 3x, \qquad y(0) = -1.$$

Solution: We begin by computing the Laplace transform of each term of the equation, obtaining

$$sY(s) - y(0) - 2Y(s) = \frac{3}{s^2}.$$

If we apply the initial condition $y(0) = -1$ and solve for $Y(s)$ we get

$$Y(s) = \frac{3}{s^2(s - 2)} - \frac{1}{s - 2}. \tag{6.1}$$

Note that the first term in the right-hand side of the above equation is the product of two known Laplace transforms, namely

$$\frac{3}{s^2(s - 2)} = 3\frac{1}{s^2} \cdot \frac{1}{s - 2} = 3\mathcal{L}\left[x\right] \cdot \mathcal{L}\left[e^{2x}\right].$$

By the convolution theorem, we know that

$$\mathcal{L}\left[x\right] \cdot \mathcal{L}\left[e^{2x}\right] = \mathcal{L}\left[x * e^2 x\right].$$

To use this, we need to compute the convolution

$$x * e^2 x = \int_0^x (x - \xi)e^{2\xi} \, d\xi.$$

Note that, because of the commutative property of the convolution, we may compute, instead, the convolution $e^{2x} * x$, which is an integral a little bit easier to evaluate. Here is what we obtain,

$$
\begin{aligned}
e^{2x} * x &= \int_0^x e^{2(x-\xi)}\xi d\xi \\
&= e^{2x}\int_0^x \xi e^{-2\xi}\, d\xi \\
&= e^{2x}\left[-\frac{1}{2}\xi e^{-2\xi} + \frac{1}{2}\left(\frac{1}{2}e^{2\xi}\right)\right]_0^\xi \\
&= e^{2x}\left(-\frac{1}{2}x e^{-2x} - \frac{1}{4}e^{-2x} + \frac{1}{4}\right) \\
&= -\frac{1}{2}x - \frac{1}{4} + \frac{1}{4}e^{2x}.
\end{aligned}
$$

Therefore, we have that

$$
\frac{3}{s^2(s-2)}\mathcal{L}\left[x\right]\mathcal{L}\left[e^{2x}\right] = 3\mathcal{L}\left[-\frac{1}{2}x - \frac{1}{4} + \frac{1}{4}e^{2x}\right],
$$

so the inverse Laplace transform of $\frac{3}{s^2(s-2)}$ is

$$
\mathcal{L}^{-1}\left[\frac{3}{s^2(s-2)}\right] = -\frac{3}{2}x - \frac{3}{4} + \frac{3}{4}e^{2x}.
$$

Using this result in **(6.1)**, we find that

$$
y(x) = -\frac{3}{2}x - \frac{3}{4} + \frac{3}{4}e^{2x} - e^{2x} = -\frac{3}{2}x - \frac{3}{4} - \frac{1}{4}e^{2x},
$$

which is the solution to our IVP. Finally, we solve this IVP using *Mathematica* to confirm our result:

```
In[1]:= Clear[y, x];
        sol = DSolve[{y'[x] - 2 y[x] == 3 x, y[0] == -1}, y[x], x];
        Set @@ sol[[1, 1]]
```

Out[1]= $\frac{1}{4}\left(-3 - e^{2x} - 6x\right)$

Multiplying out we see that it's exactly the same that we got

```
In[2]:= Expand[y[x]]
```

Out[2]= $-\frac{3}{4} - \frac{e^{2x}}{4} - \frac{3x}{2}$

6.1.3 Step Functions

Now we will consider problems in which the forcing function $g(x)$ involves a step function. Solving this kind of problem is one of the most useful elementary applications of the Laplace transform. The process of solution is just the same as before, but now we will be making use of the Laplace transform of the unit step function, as well as of the translation principle of the Laplace transform. Let's look at an example.

Example 6.5

Solve the initial value problem

$$y'' - y' - 2y = u_2(x) + 1, \qquad y(0) = 0, \ y'(0) = 0.$$

Solution: Just as before, we compute the Laplace transform of each term in the equation, obtaining

$$s^2 Y(s) - sy(0) - y'(0) - sY(s) + y(0) - 2Y(s) = \frac{e^{-2s}}{s} + \frac{1}{s}.$$

Applying the initial conditions and factoring $Y(s)$, we get

$$Y(s)(s^2 - s - 2) = \frac{e^{-2s} + 1}{s},$$

so

$$Y(s) = \frac{e^{-2s} + 1}{(s^2 - s - 2)} = \frac{e^{-2s} + 1}{s(s+1)(s-2)}. \tag{6.2}$$

Now we have to write the right-hand side in terms of simpler fractions. To do the partial fraction decomposition, it is convenient to write

$$\frac{e^{-2s} + 1}{s(s+1)(s-2)} = (e^{-2s} + 1)\left(\frac{1}{s(s+1)(s-2)}\right),$$

and then compute the partial fraction decomposition of $\frac{1}{(s+1)(s-2)s}$. First we write

$$\frac{1}{s(s+1)(s-2)} = \frac{A}{s} + \frac{B}{s+2} + \frac{C}{s-2},$$

and then we must have

$$A(s+1)(s-2) + Bs(s-2) + Cs(s+1) = 1.$$

Multiplying out and collecting terms of powers of s we find that

$$(A + B + C)s^2 + (-A - 2B + C)s - 2A = 1.$$

Comparing the coefficients of like powers of s in the LHS and the RHS we see that we must have

$$A + B + C = 0, \qquad -A - 2B + C = 0, \qquad -2A = 1,$$

so $A = -\frac{1}{2}$, $B = \frac{1}{3}$ and $C = \frac{1}{6}$. Hence, we can rewrite **(6.2)** as

$$Y(s) = (e^{-2s} + 1)\left(-\frac{1}{2s} + \frac{1}{3(s+1)} + \frac{1}{6(s-2)}\right).$$

Multiplying out in the right-hand side and pulling out the constants, we get

$$Y(s) = -\frac{1}{2}e^{-2s}\left(\frac{1}{s}\right) + \frac{1}{3}e^{-2s}\left(\frac{1}{s+1}\right) + \frac{1}{6}e^{-2s}\left(\frac{1}{s-2}\right)$$
$$-\frac{1}{2}\left(\frac{1}{s}\right) + \frac{1}{3}\left(\frac{1}{s+1}\right) + \frac{1}{6}\left(\frac{1}{s-2}\right).$$

Computing the inverse Laplace transform of functions like the last three terms is something we have been doing, but the first three terms are different. For the first term, we recognize from the table of Laplace transforms that

$$\mathcal{L}^{-1}\left[-\frac{1}{2}\frac{e^{-2s}}{s}\right] = -\frac{1}{2}u_2(x).$$

To compute the inverse Laplace transform of the other two remaining terms we have to use the translation property:

$$\mathcal{L}\left[u_c(x)f(x-c)\right] = e^{-cs}F(s) = e^{-cs}\mathcal{L}\left[f(x)\right].$$

This can also be written in terms of the inverse Laplace transform:

$$\mathcal{L}^{-1}\left[e^{-cs}F(s)\right] = u_c(x)f(x-c).$$

Let's apply this property to find the inverse Laplace transform of the second term, $\frac{1}{3}e^{-2s}\left(\frac{1}{s+1}\right)$. Note that

$$e^{-2s}\frac{1}{s+1} = e^{-2s}\mathcal{L}\left[e^{-x}\right],$$

so if we set $f(x) = e^{-x}$ and $c = 2$, the translation property will tell us that

$$\mathcal{L}^{-1}\left[e^{-2s}\mathcal{L}\left[e^{-x}\right]\right] = u_2(x)f(x-2) = u_2(x)e^{-(x-2)} = u_2(x)e^{-x+2}.$$

Hence, the inverse Laplace transform of the second term is

$$\frac{1}{3}u_2(x)e^{-x+2}.$$

Similarly, we find that

$$\mathcal{L}^{-1}\left[\frac{1}{6}e^{-2s}\frac{1}{s-2}\right] = \frac{1}{6}u_2(x)e^{2(x-2)}.$$

The computation of the remaining terms is not different from what we have done in the previous examples, and it is left to you to verify that the final solution is

$$y(x) = -\frac{1}{2}u_2(x) + \frac{1}{3}u_2(x)e^{-x+2} + \frac{1}{6}u_2(x)e^{2x-4} - \frac{1}{2} + \frac{1}{3}e^{-x} + \frac{1}{6}e^{2x}.$$

Example 6.6

Solve the IVP

$$y'' - 5y' + 6y = -u_\pi(x)\sin(x), \qquad y(0) = 0,\ y'(0) = 0. \tag{6.3}$$

Solution: This time we will use *Mathematica* to do the computations for us. Note that there are several *Mathematica* functions that can shorten the process of solution. However, we will try to solve this problem going through the details, without using the *Mathematica* functions that will actually take us to the solution in big steps. The purpose, remember, is not only to show you how to do specific algebraic manipulations with *Mathematica*, but also to allow you to solve problems by hand and verify your results at each step using this software.

Let's begin by giving a name in *Mathematica* to our equation **(6.3)**:

In[1]:= `Clear[Y, y, DiffEq, LTeq, sol, fsol];`
`DiffEq = (y')'[x] - 5 y'[x] + 6 y[x] == -UnitStep[x - π] Sin[x]`

Out[1]:= `6 y[x] - 5 y'[x] + y''[x] == -Sin[x] UnitStep[-π + x]`

Now we take the Laplace transform of this DE, giving the name **LTeq** to the result:

In[2]:= `LTeq = LaplaceTransform[DiffEq, x, s]`

Out[2]:= `6 LaplaceTransform[y[x], x, s] + s² LaplaceTransform[y[x], x, s] -`
`5 (s LaplaceTransform[y[x], x, s] - y[0]) - s y[0] - y'[0] ==` $\frac{e^{-\pi s}}{1+s^2}$

This result is too large, and we can read it better with the following replacement:

In[3]:= `LTeq = LTeq /. {LaplaceTransform[y[x], x, s] → Y[s]}`

Out[3]:= `-s y[0] + 6 Y[s] + s² Y[s] - 5 (-y[0] + s Y[s]) - y'[0] ==` $\frac{e^{-\pi s}}{1+s^2}$

Now we apply the initial conditions, again by doing a replacement:

In[4]:= `LTeq = LTeq /. {y[0] → 0, y'[0] → 0}`

Out[4]:= `6 Y[s] - 5 s Y[s] + s² Y[s] ==` $\frac{e^{-\pi s}}{1+s^2}$

We can use the function **DSolve** to solve for $Y(s)$. Remember that this will return the solution as a rule like **Y[s]->exp**, where **exp** is the expression defining the solution. Here is exactly what we obtain

In[5]:= `sol = Solve[LTeq, Y[s]]`

Out[5]:= $\left\{\left\{Y[s] \to \frac{e^{-\pi s}}{(1+s^2)(6-5s+s^2)}\right\}\right\}$

Instead of the above rule, let's redefine **sol** as the actual solution with the following replacement (since the solution is inside a list, we also extract it by taking the first part of this list).

In[6]:= **sol = (Y[s] /. sol) [[1]]**

Out[6]:= $\dfrac{e^{-\pi s}}{\left(1+s^2\right)\,\left(6-5\,s+s^2\right)}$

Now we use the function **Apart** to apply partial fraction decomposition on this result. We will also expand the result to

In[7]:= **Apart[sol, s]**

Out[7]:= $\dfrac{e^{-\pi s}}{10\,(-3+s)} - \dfrac{e^{-\pi s}}{5\,(-2+s)} + \dfrac{e^{-\pi s}\,(1+s)}{10\,\left(1+s^2\right)}$

The last fraction in this result can be split into two fractions:

In[8]:= **Expand[Apart[sol, s]]**

Out[8]:= $\dfrac{e^{-\pi s}}{10\,(-3+s)} - \dfrac{e^{-\pi s}}{5\,(-2+s)} + \dfrac{e^{-\pi s}}{10\,\left(1+s^2\right)} + \dfrac{e^{-\pi s}\,s}{10\,\left(1+s^2\right)}$

Now you must refer to the table of Laplace transforms to get the inverse Laplace transform of each fraction. Note that the presence of $e^{-\pi s}$ in each fraction tells you that you must use the translation property. Once you get your answer, compare it to the following result given by *Mathematica*:

In[9]:= **fsol = Expand[InverseLaplaceTransform[Apart[sol, s], s, x]]**

Out[9]:= $-\dfrac{1}{5}\,e^{-2\,\pi+2\,x}\,\text{HeavisideTheta}[-\pi+x] +$
$\dfrac{1}{10}\,e^{-3\,\pi+3\,x}\,\text{HeavisideTheta}[-\pi+x] -$
$\dfrac{1}{10}\,\text{Cos}[x]\,\text{HeavisideTheta}[-\pi+x] - \dfrac{1}{10}\,\text{HeavisideTheta}[-\pi+x]\,\text{Sin}[x]$

Here, *Mathematica* is using the function **HeavisideTheta** instead of **UnitStep**, which is the one we started with. Even though these are different functions, here we can ignore their difference and think of **HeavisideTheta** as the unit step function.

Exercises

(1) Use the Laplace transform and the inverse Laplace transform to solve the following initial value problems problems.

   (a) $y' = -ky$, $y(0) = 1$.

   (b) $y'' + 4y' + 3y = 0$, $y(0) = 3$, $y'(0) = 1$.

   (c) $y' = ay + b$, $y(0) = k$, with k, a and b constants.

   (d) $y'' - 10y' + 9y = t$, $y(0) = -1$, $y'(0) = 2$.

   (e) $2y'' + 3y' - 2y = te^{-2t}$, $y(0) = 0, y'(0) = -20$.

   (f) $y'' + 4y' + 3y = \sin x$, $y(0) = 1$, $y'(0) = 3$.

   (g) $y'' + ty' - 2y = 2$, $y(0) = 0$, $y'(0) = 0$.

   (h) $y'' - 6y' + 8y = u_3(x)$, $y(0) = 0$, $y'(0) = 1$.

(i) $y'' - y' = \cos t + \cos(t - 6)u_4(t)$, $y(0) = -1$, $y'(0) = 0$.

(j) $y'' + 3y' + 2y = F(t)$, $y(0) = 0$, $y'(0) = -1$, where $F(t)$ is the piecewise function

$$F(t) = \begin{cases} 1, & t < 6; \\ t, & 6 \le t < 9 \\ ;3, & t \ge 9. \end{cases}$$

(k) $y'' + 2y' - 15y = \delta(t - 4)$, $y(0) = -1$, $y'(0) = 5$.

(2) **Challenge Problems.**

(a) Let $f(t)$ be a piecewise continuous function on every interval where $t \ge 0$. The function $f(t)$ also satisfies the following property

$$|f(t)| \le Me^{kt} \qquad \text{for all } t \ge .$$

where K and M are constants. Show that $\mathcal{L}[f(t)](s)$ exists for all $s > K$.

(b) Show that $\lim_{t \to \infty} f(t) = \lim_{s \to \infty} s\mathcal{L}[f(t)](s)$.

(c) Assuming the definition of the inverse Laplace transform to be

$$f(t) = \mathcal{L}^{-1}[F(s)] = \frac{1}{2\pi i} \int_{\gamma - i\infty}^{\gamma + i\infty} e^{st} F(s)\, ds,$$

prove that

$$f(0^+) = \lim_{s \to \infty} sF(s) \qquad \text{and} \qquad f(\infty) = \lim_{s \to 0} sF(s).$$

(d) Use the formula $\mathcal{L}[y'] = sY(s) - y(0)$ to derive a formula for

$$\mathcal{L}\left[\int_0^x f(t)\, dt\right].$$

(e) Use the facts

$$\mathcal{L}\left[\frac{f(x)}{x}\right] = \int_s^\infty F(s)\, ds \qquad \text{and} \qquad \mathcal{L}[\sin x] = \frac{1}{s^2 + 1}$$

to prove that

$$\int_0^\infty \frac{\sin x}{x}\, dx = \frac{\pi}{2}.$$

(f) Suppose $f(x)$ is a periodic function with period P, that is,

$$f(x + P) = f(x)$$

for all x in the domain of f. Find an expression for

$$\mathcal{L}[f(x)].$$

(g) Consider

$$L\frac{dI}{dt} + RI = E, \qquad I(0) = I_0.$$

Use the Laplace transform technique to solve this IVP if

    i. $E = E_0 u(t)$.

    ii. $E = E_0 \delta(t)$.

(h) Consider

$$L\frac{d^2 I}{dt^2} + R\frac{dI}{dt} + \frac{1}{C}I = g(t), \qquad I(0) = I_0, \ I'(0) = 0,$$

and

$$g(t) = \begin{cases} -10, & 0 < t < \pi; \\ 0, & \pi < t < 4\pi; \\ 10, & t > 4\pi. \end{cases}$$

Using the Laplace transform, solve for $I(t)$. (Imagine solving this by using conventional methods. This problem, hopefully, will help you appreciate the method of Laplace transform.)

(3) **Application of the Laplace Transform**

Note: Laplace transforms have been needed in several areas, such as circuit theory, problems in economics and finance, physics and engineering. This section is dedicated to such problems. Use the Laplace transform and the inverse Laplace transform to solve the DEs arising in these problems.

(a) Harmonic vibrations of a beam supported at two ends:
For a beam of length l whose normal deflection $w(x, t)$ is measured downward if the axis of the beam is toward the x-axis. The equation of motion is given by

$$EI\frac{d^4 w}{dx^4} - m\omega_0^2 w = 0,$$

where E is the Young modulus of elasticity, I is the moment of inertia, m the mass per unit length and ω_0 is the angular frequency. Use the Laplace transform and $F(0) = 0$, $F(l) = 0$, $F''(0) = 0$ and $F''(l) = 0$ to solve the DE.

(b) Electric circuits:
Consider a circuit with a resistor R, inductor L and capacitor C. The current flow in the circuit is governed by the DE

$$L\frac{d^2 I}{dt} + \frac{1}{C}I + R\frac{dI}{dt} = 0.$$

Assuming appropriate initial conditions, solve this DE using the Laplace transform technique.

(c) GDP problem from economics:
The rate of change of GDP is proportional to the current GDP. This is modeled as first order DE:

$$x'(t) = gx(t),$$

where $g = \frac{x'}{x}$ is the grow rate and t stands for time. Assume $x(0) = k$, and use the Laplace transform to solve the IVP. What happens when $g < 0$?

(d) Newton's law of cooling:
The following DE describes the change in the temperature of an object in a given environment. The law states that the rate of change of the temperature is proportional to the difference between the temperature of the object and the temperature of the environment:

$$\frac{dT}{dt} = -K(T - T_e).$$

Assume that the temperature at time $t = 0$ is T_0. Use the Laplace transform to solve this IVP. Explain the solution.

Chapter 7

A Short Discussion on Power Series Solutions to DEs

The author intends to keep this chapter short. The pointer is provided for further reading in the case some students are extremely motivated. It is advised that this chapter is better taught in a standard one semester/quarter long course on ordinary differential equations. Only one example problem is solved and no exercise problems are provided. Further reading is suggested at the end of the chapter.

7.1   Review of Power Series

In ordinary DEs, the power series method is used to seek solutions. In general, such solutions assume the form of a power series with unknown coefficients. The coefficients are determined by using a **recurrence relation**.

A power series about a is a series that can be written in the form

$$\sum_{n=0}^{\infty} c_n (x-a)^n, \qquad \text{where } a \text{ and } c_n \text{ are constants.}$$

The c_n are the unknown coefficients.

Note:

(1) A power series is a function of x.

(2) Convergence is a very important issue that has to be dealt with. A power series may or may not converge depending on the values of x.

159

Definition 7.1

There is always a number r such that the power series will converge for $|x-a| < r$ and diverge for $|x-a| > r$. This number r is called the radius of convergence of the power series.

Note:

(1) The points $|x - a| = R$ will not change the radius of convergence.

(2) We may find that for some power series the radius of convergence is $r = \infty$. This means that the power series will converge for any value of x, i.e. the series will always converge.

Definition 7.2

The information about the radius of convergence of a power series defines the **interval of convergence** of the same, i.e. if r is the radius of convergence, the power series converges for all x satisfying

$$a - r < x < a + r.$$

Thus, the interval of convergence is $(a - r, a + r)$. Plus, if the series also converges for $x = a - r$ and/or for $x = a + r$, we need to include those points as well in the interval of convergence.

> **Remark**
> The power series always converges for $x = a$.

Example 7.3

Determine the radius of convergence and interval of convergence of the power series

$$\sum_{n=1}^{\infty} \frac{n}{4^n}(x + 2)^n.$$

Solution: Note: the power series converges at $x = 2$. In order to find other values of x for which the series converges, we need to employ some test for series convergence (in this case, the **ratio test**):

$$
\begin{aligned}
L &= \lim_{n \to \infty} \left| \frac{\frac{(n + 1)(x + 2)^{n+1}}{4^{n+1}}}{\frac{n(x + 2)^n}{4^n}} \right| \\
&= \lim_{n \to \infty} \left| \frac{n + 1}{n} \cdot \frac{4^n}{4^{n+1}}(x + 2) \right| \\
&= \lim_{n \to \infty} \left| \frac{n + 1}{n} \right| \frac{|x + 2|}{4} \\
&= 1 \cdot \frac{|x + 2|}{4} \\
&= \frac{|x + 2|}{4}.
\end{aligned}
$$

According to the ratio test, the series will converge if $L < 1$, that is, if

$$\frac{|x+2|}{4} < 1 \quad \Rightarrow \quad |x+2| < 4 \quad \Rightarrow \quad -6 < x < 2.$$

Hence, the radius of convergence is $r = 4$. To determine the interval of convergence, we also need to check the endpoints $x = -6$ and $x = 2$. For $x = -6$, we have

$$\sum_{n=1}^{\infty} \frac{n}{4^n}(-4)^n = \sum_{n=1}^{\infty} \frac{n}{4^n}(-1)^n(4)^n$$
$$= \sum_{n=1}^{\infty} (-1)^n n,$$

which is an alternating series that diverges. Now let's use $x = 2$:

$$\sum_{n=1}^{\infty} \frac{n}{4^n}(4)^n = \sum_{n=1}^{\infty} n,$$

which diverges since $\lim_{n \to \infty} = \infty \neq 0$. Hence, the interval of convergence is

$$-6 < x < 2.$$

Note: For a range of values of x, certain functions can be represented as a power series

$$f(x) = \sum_{n=0}^{\infty} c_n(x-a)^n = c_0 + c_1(x-a) + c_2(x-a)^2 + \cdots$$

7.2 Differentiation and Integration of Power Series

If $f(x) = \sum_{n=0}^{\infty} c_n(x-a)^n$ has a radius of convergence $r > 0$, then

$$f'(x) = \sum_{n=0}^{\infty} nc_n(x-a)^{n-1}$$

and

$$\int f(x) = C + \sum_{n=0}^{\infty} \frac{(x-a)^{n+1}}{n+1}.$$

Both $f'(x)$ and $\int f(x)$ will also have radius of convergence r
Facts and tricks required:

Fact 1: If

$$\sum_{n=0}^{\infty} a_n(x - x_0)^n = 0,$$

then a_n is identically zero for all values of $n = 1, 2, 3\ldots$.

Fact 2: If $f(x)$ can be differentiated infinitely, then the Taylor series of $f(x)$ about $x = a$ is

$$f(x) = \sum_{n=0}^{\infty} \frac{f^{(n)}(a)}{n!}(x - a)^n,$$

where $f^{(n)}(a)$ is the n-th derivative of $f(x)$ at $x = a$.

Example 7.4

Which is the power series representation of e^x and $\cos x$ about 0.

Solution: In calculus, you were shown that

$$e^x = \sum_{n=0}^{\infty} \frac{x^n}{n!}$$

and

$$\cos x = \sum_{n=0}^{\infty} \frac{(-1)^n x^{2n}}{(2n)!}.$$

Definition 7.5

A function f is called analytic at $x = a$ if the Taylor series for $f(x)$ converges to $f(x)$ and has a radius of convergence $r > 0$.

Trick (Index Shift): Let

$$f(x) = \sum_{n=0}^{\infty} c_n(x - a)^n = c_0 + c_1(x - a) + c_2(x - a)^2 + \cdots + c_n(x - a)^n + \cdots,$$

and

$$f'(x) = \sum_{n=0}^{\infty} nc_n(x-a)^{n-1} = c_1 + 2c_2(x-a) + 3c_3(x-a)^2 + \cdots + nc_n(x-a)^{n-1} + \cdots,$$

so we see that $f'(x)$ starts at c_1. In order to have a series representation of $f'(x)$ we rewrite $f'(x)$ as

$$f'(x) = 1c_1(x - a)^0 + 2c_2(x - a)^1 + 3c_3(x - a)^2 + \cdots + nc_n(x - a)^{n-1} + \cdots,$$

that is,

$$f'(x) = \sum_{n=1}^{\infty} nc_n(x - a)^n$$

But, if the series representation has to begin from $n = 0$, just like $f(x)$, what are we supposed to do? In this case, it is easy, since $n = 0$ will make the first term equal to zero, and therefore we can make

$$f'(x) = 0 + \sum_{n=0}^{\infty} nc_n(x-a)^{n-1} = \sum_{n=0}^{\infty} nc_n(x-a)^{n-1}.$$

This is called **index shift**. The technique is particularly useful in seeking power series solutions to a DE because $f'(x)$ and $f''(x)$ need to be computed and the indices arranged in the same way. This will help us to use fact 1 if needed.

7.3 Solving DEs using Power Series

Consider the DE

$$p(x)y'' + q(x)y' + r(x)y = 0, \qquad p(x) \neq 0, \tag{7.1}$$

which can be written as

$$y'' + \frac{q(x)}{p(x)}y' + \frac{r(x)}{p(x)} = 0.$$

In this section, we will discuss ordinary points. A point $x = a$ is an **ordinary point** if $\frac{q(x)}{p(x)}$ and $\frac{r(x)}{p(x)}$ are analytic at $x = a$ (see definition 7.5). Since $p(a) \neq 0$ by assumption, the problem becomes much nicer, and the point a becomes an ordinary point.

Algorithm

(1) Assume $y(x) = \sum_{n=0}^{\infty} c_n(x-a)^n$ is a solution to **(7.1)**.

(2) Evaluate $y'(x)$ and $y''(x)$ and plug-in the equation.

(3) Use the index shift to have all series start at the same value of n.

(4) Combine all the series into a single one.

(5) Use fact 1 to obtain a relation in terms of c_n (which will be a recurrence relation).

(6) Obtain a representation for c_n.

(7) Finally, obtain $y(x)$.

Example 7.6

Determine a series solution to

$$y'' + y = 0$$

about $a = 0$.

Solution: Note, $p(x) = 1$, so we are dealing with ordinary point here. We seek a solution

$$y(x) = \sum_{n=0}^{\infty} c_n x^n.$$

First, we find the derivatives

$$y'(x) = \sum_{n=0}^{\infty} n c_n x^{n-1}$$

and

$$y''(x) = \sum_{n=2}^{\infty} n(n-1) c_n x^{n-2}. \qquad \text{(easy check!)}$$

Hence, the DE becomes

$$\sum_{n=0}^{\infty} n(n-1) c_n x^{n-2} + \sum_{n=0}^{\infty} c_n x^n = 0.$$

Now we need to perform the index shift: we need to shift the first power series from $n = 2$ to $n = 0$:

$$\sum_{n=2}^{\infty} n(n-1) c_n x^{n-2} = \sum_{n=0}^{\infty} (n+2)(n+1) c_{n+2} x^n.$$

It is advisable to get the exponent as n. A little practice is needed.

Now the DE becomes

$$\sum_{n=0}^{\infty} (n+2)(n+1) c_{n+2} x^n + \sum_{n=0}^{\infty} c_n x^n = 0,$$

and combining the two series into a single one, we get

$$\sum_{n=0}^{\infty} \left[(n+2)(n+1) c_{n+2} c_n\right] x^n = 0.$$

Now recall fact 1; this gives us

$$(n+2)(n+1) c_{n+2} + c_n = 0.$$

Solving for c_{n+2}, we get

$$c_{n+2} = -\frac{c_n}{(n+2)(n+1)}, \qquad n = 0, 1, 2, \ldots$$

Let's evaluate this expression for some values of n.

$$n = 0 \qquad c_2 = -\frac{c_0}{2 \cdot 1}$$

$$n = 2 \qquad c_4 = -\frac{c_2}{4 \cdot 3} = \frac{c_0}{4 \cdot 3 \cdot 2 \cdot 1}$$

$$n = 4 \qquad c_6 = -\frac{c_4}{6 \cdot 5} = -\frac{c_0}{6!}.$$

We see that the pattern is

$$c_{2k} = \frac{(-1)^k}{(2k)!} c_0, \qquad k = 1, 2, 3, \ldots$$

Also, we have

$$n = 1 \qquad c_3 = -\frac{c_1}{3 \cdot 2}$$

$$n = 3 \qquad c_5 = -\frac{c_3}{5 \cdot 4} = \frac{c_1}{5 \cdot 4 \cdot 3 \cdot 2 \cdot 1},$$

so

$$c_{2k+1} = \frac{(-1)^k}{(2k+1)!} c_1, \qquad k = 1, 2, 3, \ldots$$

Now, the solution,

$$
\begin{aligned}
y(x) &= \sum_{n=0}^{\infty} c_n x^n \\
&= c_0 + c_1 x + c_2 x^2 + c_3 x^3 + \cdots + c_{2k} x^{2k} + c_{2k+1} x^{2k+1} + \cdots \\
&= c_0 + c_1 x - \frac{c_0}{2!} x^2 - \frac{c_1}{3!} x^3 + \cdots + \frac{(-1)^k}{(2k)!} c_0 x^{2k} + \frac{(-1)^{k+1} c_1}{(2k+1)!} x^{2k+1} + \cdots
\end{aligned}
$$

Collecting the terms of c_0 and c_1, we have

$$
\begin{aligned}
y(x) &= c_0 \left[1 - \frac{1}{2!} x^2 + \cdots + \frac{(-1)^k}{(2k)!} x^{2k} + \cdots \right] \\
&\quad + c_1 \left[x - \frac{1}{3!} x^3 + \cdots + \frac{(-1)^{k+1}}{(2k+1)!} x^{2k+1} + \cdots \right] \\
&= c_0 \sum_{n=0}^{\infty} \frac{(-1)^k}{(2k)!} x^{2k} + c_1 \sum_{n=0}^{\infty} \frac{(-1)^k}{(2k+1)!} x^{2k+1} \\
&= c_0 \cos x + c_1 \sin x,
\end{aligned}
$$

where the last equality comes from the Taylor series representation for $\sin x$ and $\cos x$. Recall that this is the solution we obtained otherwise in Chapter 3.

Suggested Texts for Further Reading

(1) Weisstein, Eric W. "Frobenius Method."

(2) Paul Dawkins. "Lamar Tutorials."

(3) Boyce and DiPrima. *Elementary Differential Equations.*

(4) Coddington and Levinson. *Theory of Ordinary Differential Equations.*

(5) Edwards and Penney. *Differential Equations and Boundary Value Problems.*

(6) Tenenbaum and Pollard. *Ordinary Differential Equations.*

Chapter 8

Introduction to *Mathematica*

In this chapter, we offer a quick introduction to *Mathematica*, a very powerful computer algebra system (CAS). The purpose of this introduction is to teach you enough about this software so that you can understand the lines of *Mathematica* code used through out the book. Since any installation of *Mathematica* includes formidable documentation, we allow ourselves to omit many details, which are introduced and thoroughly explained in *Mathematica*'s own documentation.

8.1 The *Mathematica*'s front end

Mathematica is composed of two major elements: the **kernel** and the **front-end**. The kernel is the part that actually performs all computations, while the front-end serves as a graphical interface through which we communicate with the kernel and pass to it the expressions that are to be evaluated.

When you open a *Mathematica* session ordinarily, what you will see is the front-end. The front end will by default organize any input and output using in the form of a *Mathematica* **notebook**. A notebook is structured as a sequence of **cells**, the basic units within which information is contained. If you start typing after initializing a *Mathematica* session, you will see a bracket to the right indicating the limits of the cell in which you are typing. To evaluate the expression you typed, press shift+enter. Figure 8.1 shows an example of what you see after evaluating 2+2 in the front-end. Note that *Mathematica* assigns a label like in[1] and out[1] to each input and output in a *Mathematica* session.

In this book, we have simulated input and output code as displayed in the *Mathematica*'s front-end, but we have omitted some characteristics such as cell brackets indicating the scope of a cell. Here is, for example, the way we represent the input and output that you see in figure 8.1:

167

Figure 8.1. Evaluating **2+2** in the front-end of *Mathematica*

In[1]:= **2 + 2**

Out[1]:= 4

8.2   First Computations with *Mathematica*

Basic arithmetic operations are done in the same way you do them in a calcula-
tor. However, there are also other ways to indicate some arithmetic operations.
For example, multiplication is indicated with an asterisk or a blank space. Here
is an example, where we assign some numerical values to **a** and **b** and then
compute their product.

In[2]:= **a = 3; b = 5; a b**

Out[2]:= 15

Equivalently, we have

In[3]:= **a b**

Out[3]:= 15

Some other operations require explicitly using a built-in *Mathematica* function,
such as **Sqrt** to compute a square root. But what we want to let you know is
that *Mathematica* offers ways to input operations using ordinary mathematical
notation. In the front-end, you can input

In[4]:= 3^2

Out[4]:= 9

or

In[5]:= $\frac{3}{15}$

Out[5]:= $\frac{1}{5}$

To do this, you can use, for example, the *Mathematica* palette "Basic Math Assistant" in the menu Palettes. There you will find buttons to insert constructions to input several mathematical operations using conventional notation. To facilitate reading, this is the way in which we will present our operations here.

By default, *Mathematica* will try to return exact results, often using symbolic representations for its output. For example, we have

In[6]:= $\sum_{n=0}^{\infty} \frac{1}{n!}$

Out[6]:= e

If our input contains a decimal point, then *Mathematica* will return a decimal approximation:

In[7]:= $\sum_{n=0}^{\infty} \frac{1.}{n!}$

Out[7]:= 2.71828

Of course, we can always ask *Mathematica* for a decimal approximation of our results. One way to do this is using the function **N**. For example,

In[8]:= **N[π]**

Out[8]:= 3.14159

Here is a more accurate approximation with 300 significant figures:

In[9]:= **N[π, 300]**

Out[9]:= 3.14159265358979323846264338327950288419716939937510582097494`
45923078164062862089986280348253421170679821480865132823066４`
70938446095505822317253594081284811174502841027019385211055５`
96446229489549303819644288109756659334461284756482337867831６`
52712019091456485669234603486104543266482133936072602491412７

Of course, *Mathematica* is able to perform not only numerical computations, but also symbolic ones. For example,

In[10]:= **Clear[a, b]**

Out[10]:=

In[11]:= **Simplify$\left[(a + b)^2 - a^2\right]$**

Out[11]:= $b\,(2\,a + b)$

This is a very basic example, since *Mathematica* is able to handle much more complicated expressions and apply a wide range of transformations to them.

We can even pass some assumptions to *Mathematica* on the nature of our symbols:

In[12]:= **Simplify[Cos[x + 2 k π], k ∈ Integers]**

Out[12]:= Cos[x]

To learn more about each function such as **Simplify**, refer to the "Documentation Center" in the menu Help.

8.3   Lists

We can say that lists are the fundamental data structures of *Mathematica*. A list of elements is produced in *Mathematica* enclosing its elements in curly brackets:

In[13]:= **a = {1, 2, 3, 4}**

Out[13]:= {1, 2, 3, 4}

There are many operations we can do on a list. For example,

In[14]:= **Reverse[a]**

Out[14]:= {4, 3, 2, 1}

Some functions with the attribute **Listable** map themselves into the elements of a list. This is the case of the basic arithmetic operations. We have, for instance,

In[15]:= $\left\{ \mathbf{a + x, \ a\,x, \ a^2, \ \sqrt{a}} \right\}$

Out[15]:= $\Big\{ \{1 + x, \ 2 + x, \ 3 + x, \ 4 + x\},$
$\{x, \ 2\,x, \ 3\,x, \ 4\,x\}, \ \{1, \ 4, \ 9, \ 16\}, \ \left\{1, \ \sqrt{2}, \ \sqrt{3}, \ 2\right\}\Big\}$

Many functions may not have this property, and we have to use the function **Map** to achieve this type of results. For example, we can map a function **funct** (so far undefined) into the elements of list **a** as follows:

In[16]:= **funct /@ a**

Out[16]:= {funct[1], funct[2], funct[3], funct[4]}

Here is a equivalent way to do this using the infix form of **Map**:

In[17]:= **funct /@ a**

Out[17]:= {funct[1], funct[2], funct[3], funct[4]}

We can also add, subtract, multiply, etc. two lists provided they are of the same length (same number of elements), since the operations are done elementwise:

In[18]:= **{1, 2, 3} + {x, y, z}**

Out[18]:= {1 + x, 2 + y, 3 + z}

or

In[19]:= $\dfrac{\{1,2,3\}}{\{x,y,z\}}$

Out[19]= $\left\{\dfrac{1}{x},\ \dfrac{2}{y},\ \dfrac{3}{z}\right\}$

This is also a result of the arithmetic functions being listable. To do something similar with an arbitrary function, you can use **Thread**:

In[20]:= **Thread[funct[{1, 2, 3, 4}, {u, v, x, y}]]**

Out[20]= {funct[1, u], funct[2, v], funct[3, x], funct[4, y]}

Another example where we produce a list of equalities that can be seen as a system of equations:

In[21]:= **Thread[{x + y, 2 x - 3 y} == {2, -1}]**

Out[21]= {x + y == 2, 2 x - 3 y == -1}

The *Mathematica* documentation center has a very good tutorial on the manipulation of lists, so at this point we advise you to refer to it for further investigation related to the use of lists in *Mathematica*.

8.4   Patterns

A very nice way to do things in *Mathematica* is through the use of **patterns**. First, let us see several ways to ask *Mathematica* to give us the parts of an expression that satisfy a certain property. Let us start by defining

In[22]:= **a = Range[100]**

Out[22]= {1, 2, 3, 4, 5, 6, 7, 8, 9, 10, 11, 12, 13, 14, 15, 16, 17, 18, 19, 20, 21, 22, 23, 24, 25, 26, 27, 28, 29, 30, 31, 32, 33, 34, 35, 36, 37, 38, 39, 40, 41, 42, 43, 44, 45, 46, 47, 48, 49, 50, 51, 52, 53, 54, 55, 56, 57, 58, 59, 60, 61, 62, 63, 64, 65, 66, 67, 68, 69, 70, 71, 72, 73, 74, 75, 76, 77, 78, 79, 80, 81, 82, 83, 84, 85, 86, 87, 88, 89, 90, 91, 92, 93, 94, 95, 96, 97, 98, 99, 100}

We can use the function **Cases** to select the numbers of **a** that satisfy a certain property. Here is, for example, one way to extract all those x of the list **a** such that $x \equiv 1 \mod 4$:

In[23]:= **Cases[a, x_ /; Mod[x, 3] == 1]**

Out[23]= {1, 4, 7, 10, 13, 16, 19, 22, 25, 28, 31, 34, 37, 40, 43, 46, 49, 52, 55, 58, 61, 64, 67, 70, 73, 76, 79, 82, 85, 88, 91, 94, 97, 100}

The expression **x_** is a pattern object that stands for any *Mathematica* expression, and **/;** is used to specify a condition, in this case **Mod[x,3]==1**. We can also make replacements, with the functions **Replace** and **ReplaceAll**, of elements of **a** of a certain type by other elements. For example, we can emphasize all the prime numbers in **a** by putting them into red color:

In[24]:= `a /. {x_ /; PrimeQ[x] :> Style[x, Red]}`

Out[24]= {1, 2, 3, 4, 5, 6, 7, 8, 9, 10, 11, 12, 13, 14, 15, 16, 17, 18, 19,
  20, 21, 22, 23, 24, 25, 26, 27, 28, 29, 30, 31, 32, 33, 34, 35, 36,
  37, 38, 39, 40, 41, 42, 43, 44, 45, 46, 47, 48, 49, 50, 51, 52,
  53, 54, 55, 56, 57, 58, 59, 60, 61, 62, 63, 64, 65, 66, 67, 68,
  69, 70, 71, 72, 73, 74, 75, 76, 77, 78, 79, 80, 81, 82, 83, 84,
  85, 86, 87, 88, 89, 90, 91, 92, 93, 94, 95, 96, 97, 98, 99, 100}

Here we are using the infix form of **ReplaceAll**, namely **/.**, which offers a shorter syntax. Here is a simple use of replacements and patterns. Imagine we have a general solution to a DE:

In[25]:= `sol = C[1] Cos[2 x] + C[2] Sin[2 x]`

Out[25]= $C[1] \, Cos[2\,x] + C[2] \, Sin[2\,x]$

If we want to remove the constants **C[1]** and **C[2]**, we replace anything of the form **C[_]** by **1**, like this,

In[26]:= `sol /. {C[_] -> 1}`

Out[26]= $Cos[2\,x] + Sin[2\,x]$

Or perhaps we want to display these constants in a different way using sub-indices. In this case, it is again useful to give a name to the pattern object being matched (the subindex):

In[27]:= `sol /. {c_x_ -> c_x}`

Out[27]= $c_2 \sin(2\,x) + c_1 \cos(2\,x)$

Replacements are just a few examples of how one can use patterns, and there is much more to them than what we can illustrate here. Again, you must refer to *Mathematica*'s own documentation for a larger set of examples and discussion on patterns. Next sections describes one more common use of them when setting user-defined functions.

8.5   Functions

Apart form simple assignments like

In[28]:= `powers = Table[2^i, {i, 0, 10}]`

Out[28]= {1, 2, 4, 8, 16, 32, 64, 128, 256, 512, 1024}

we can also define our own functions that return a value according to arguments passed to them. To do this, we again make use of patterns to name what is usually a formal parameter. Here, for example, a version of **list** defined above that can take an argument whose value determines the last power of 2 to be in the list:

In[29]:= **Clear[powers]; powers[n_] := Table$\left[2^i, \{i, 0, n\}\right]$**

Out[29]:=

Now, for example, we have

In[30]:= **powers[5]**

Out[30]:= {1, 2, 4, 8, 16, 32}

We can have more than one parameter. For example, we can define

In[31]:= **powers[b_, n_] := Table$\left[b^i, \{i, 0, n\}\right]$**

Out[31]:=

where now we can specify the base of each power, so for example we have

In[32]:= **powers[3, 5]**

Out[32]:= {1, 3, 9, 27, 81, 243}

Again, this are just the basics about defining functions in *Mathematica*, so you must refer to other source, like the *Mathematica*'s documentation, to study this more deeply.

8.6   Graphics

Mathematica has a many functions to produce graphics. We will start by looking at some that are used to produce the graph of functions in two dimensions.

To graph a function $y = f(x)$, we can use the function **Plot**. Here is an example where we plot $y = \sin x$ in the interval $[0, 2\pi]$.

In[33]:= **Plot[Sin[x], {x, 0, 2 π}]**

Out[33]:=

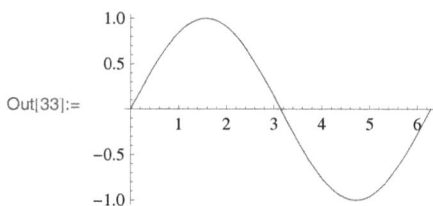

We can have the graph of more than one function drawn at once:

In[34]:= `Plot[{Sin[x], Cos[x]}, {x, 0, 2 π}]`

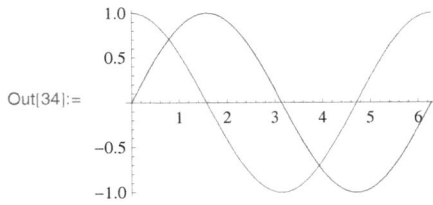

Out[34]:=

A discrete version of this function is `DiscretePlot`. For example, we can plot $y = \sin x$ in the interval $[0, 2\pi]$ using steps of size 0.1:

In[35]:= `DiscretePlot[Sin[x], {x, 0, 2 π, 0.1}]`

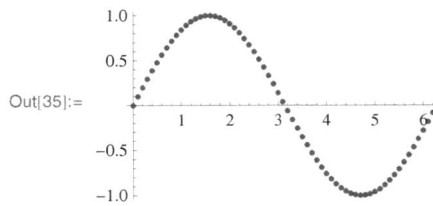

Out[35]:=

We also have the 3D version of the function `Plot`:

In[36]:= `Plot3D[x² - y², {x, -2, 2}, {y, -2, 2}]`

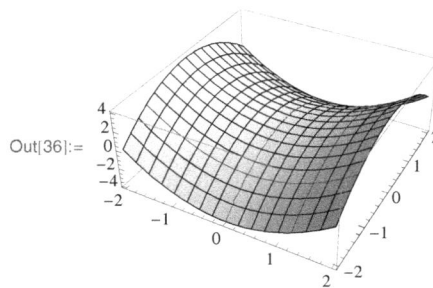

Out[36]:=

To plot the solution to a system of differential equations, we need to produce the graph of a parametrized curve given by $(x(t), y(t))$. To do this, we use *Mathematica* function `ParametricPlot`. For example,

In[37]:= `ParametricPlot[{Cos[x], Sin[3 x] Cos[x]}, {x, 0, 2 π}]`

Out[37]:=

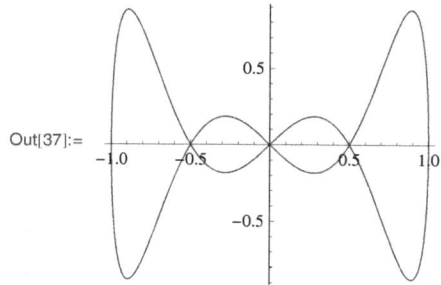

Bibliography

[1] A. A. Andronov, S. E. Khaikin, and A. A. Vitt, *Theory of Oscillators*, Dover.

[2] D. A. Sanchez, *Ordinary Differential Equations and Stability Theory: An Introduction*, Dover.

[3] E. A. Coddington, *An Introduction to Ordinary Differential Equations*, Prentice Hall.

[4] E. L. Ince, *Ordinary Differential Equations*, Dover.

[5] F. Brauer and J. A. Nohel, *Ordinary Differential Equations*, Benjamin.

[6] F. Brauer and J. A. Nohel, *Qualitative Theory of Ordinary Differential Equations*, Benjamin.

[7] G. Birkhoff and G. Rota, *Ordinary Differential Equations*, Wiley.

[8] G. F. Carrier and C. E. Pearson, *Ordinary Differential Equations*, Blaisdell.

[9] G. F. Simmons, *Differential Equations with Applications and Historical Notes*, McGraw-Hill.

[10] I. Percival and D. Richards, *Introduction to Dynamics*, Cambridge University Press.

[11] L. S. Pontryagin, *Ordinary Differential Equations*, Dover.

[12] M. W. Hirsch and S. Smale, *Differential Equations, Dynamical Systems, and Linear Algebra*, Academic Press.

[13] R. A. Struble, *Nonlinear Differential Equations*, McGraw-Hill.

[14] R. Courant and D. Hilbert, *Methods of Mathematical Physics*, Part I, Wiley-Interscience.

[15] V. I. Arnold, *Ordinary Differential Equations*, MIT Press or Springer Verlag.

[16] W. E Boyce and R. C. DiPrima, *Elementary Differential Equations and Boundary-Value Problems*, Wiley.

[17] W. Walter, *Ordinary Differential Equations*, Springer-Verlag.

Answers to Recapitulation Problems on Chapter II

(1) Homogeneous. Use a substitution $y = vx$ where v is a function of x, and then you have $y' = v'x + v$. Substitute these into the original DE and change it in terms of v and x and solve for $v(x)$. Finally, $y = vx$ will be the solution to the original DE. Don't forget to find the value of the constant of integration!

(2) Exact. Check for exactness and proceed by choosing the magic function $\varphi(x, y)$ satisfying the conditions.

(3) Be careful, it is not exact although many may be tempted to believe that! This is rather simple: a separable DE. Group the x's and the y's to integrate and obtain the solution.

(4) Comparing with the standard form, we see that the DE is linear, so it is solved using an integrating factor:

$$\mu(x) = e^{\int p(x)\, dx}.$$

In this case, $p(x) = -\frac{4}{x}$. Figure out the exact differential and follow the algorithm to finish solving the DE. TEASER: Is the DE defined at $x = 0$?

(5) Not linear. Why? It is a Bernoulli equation with $n = 3$. The substitution used to solve this equation is $v = y^{1-3} = y^{-2}$. Proceed to finish the problem.

Index

www.ingramcontent.com/pod-product-compliance
Lightning Source LLC
Chambersburg PA
CBHW081528220326
41598CB00036B/6363